Pollen:

Illustrations and Scanning Electronmicrographs

Yozo IWANAMI Tetsuo SASAKUMA Yoshio YAMADA

Pollen:

Illustrations and Scanning Electronmicrographs

With 59 Illustrations and 302 Plates

KODANSHA
Tokyo

SPRINGER-VERLAG
Berlin Heidelberg New York
London Paris Tokyo

Yozo Iwanami Dr. of Science, Plant Physiologist, Professor Emeritus of Yokohama City University, *Yokohama, Kanagawa 232, Japan*

Tetsuo Sasakuma PhD. Plant Geneticist, Associate Professor of Kihara Institute for Biological Research, Yokohama City University, *Yokohama, Kanagawa 232, Japan*

Yoshio Yamada Dr. of Science, Plant Morphologist, Professor of Department of Biology, Faculty of Education, Gunma University, *Maebashi, Gunma 371, Japan*

KODANSHA SCIENTIFIC BOOKS

ISBN3-540-18833-9
Springer-Verlag Berlin Heidelberg New York
ISBN0-387-18833-9
Springer-Verlag New York Berlin Heidelberg

ISBN4-06-202690-2(JAPAN)

Sole distribution rights outside Japan granted to
Springer-Verlag Berlin Heidelberg New York London Paris Tokyo

Copublished by

KODANSHA LTD.
12–21 Otowa 2-chome, Bunkyo-ku, Tokyo 112, Japan

and

SPRINGER-VERLAG
Tiergartenstr. 17, D-6900 Heidelberg 1, FRG

Printed in Japan

Preface

Pollen is the male gamete of the flowering plants through which genetic information is transmitted from the male parent to the offspring.

To the naked eye, pollen of flowering plants all appear to be simple yellow grains of powder. However, under the microscope, especially the scanning electron microscope, each grain possesses a species-specific pattern on its surface by which we can identify and classify the plant species. The patterns vary greatly depending on the species, exhibiting the beautiful art work of nature.

In textbooks, it is generally written that the nuclei of pollen and egg fuse to form the fertilized egg from which the embryo develops. This description may confuse students into believing that the pollen of flowering plants is similar to the sperm of animals. However, unlike sperm, pollen itself does not fuse with the egg, but two cells in a pollen grain, called sperm cells, which are differentiated from a generative cell, do undergo fusion during double fertilization, while the other cell in a pollen grain, called the vegetative cell, functions to regulate fertilization (see illustration below). In other words, a pollen grain is a complex structure consisting of a large vegetative cell within which a generative cell is included.

In plants, because these small cells are not equipped for movement, pollen grains migrate from the anthers of the stamen to the stigma aided by wind or insects. On the stigma, pollen initiates germination by forming a pollen tube which penetrates into the pistil. Through the elongated pollen tube, the sperm cells are transported to the ovule; one fuses with the egg cell while the other fuses with polar nuclei.

Thus each yellow pollen grain is the container of generative or sperm cells, and the pollen tube is the channel of transport for the these cells.

Size and shape of pollen grains vary widely according to plant species: from 10 to 200 μm in diameter with spherical, oval, cubic, hexahedral, fibrous, sicus and other shapes. Each grain has a specific pattern of fine structure on its surface.

Since the average size of a pollen grain is about 30 μm in diameter, the fine structure on the surface cannot be observed under the light microscope, which shows up only traces or furrorw. Thus, under low magnification, the pollen grain of the rice plant looks like a small vessel with a smooth surface. Under higher magnification, black spots can be identified on the surface surrounded by a network structure. However, it is not clear under the light microscope whether the spots are concave or convex. According to light microscopic observations in ordinary textbooks, the pollen grain of rice is classified as belonging to a group with a thin net structure surface. However, under a scanning electron microscope (SEM), the black spots are clearly demonstrated as granulate concaves and plugs are identified at the aperture opposite the end from which the pollen tube elongates, not only in pollen grains of rice plants but also in those of *Gramineae, Caryophyllaceae, Chenopodiaceae* and other species.

Therefore, it is necessary to reconsider pollen in light of observations using SEM, with its fine resolution power and large focus depth. The authors wish to re-establish the morphological study of pollen taking into account newly accumulated phylogenetic information.

We have been collecting SEM micrographs of various kinds of pollen over the past five years. It will take time to cover all the major species in the micrograph list. Some of these photographs are included in this volume. The readers will undoubtedly be impressed by the complexity of shapes and structures of pollen grain, which are beyond description.

This atlas was compiled using illustrations and photographs obtained in the course of our work to introduce the minute and mysterious world of pollen to as many readers as possible. The descriptions in the text closely follow our experiments and observations. We hope this volume will help broaden the understanding of the nature of pollen. It is our fervent hope that our readers will be as amazed as we are with this aspect of the beauty of nature.

The SEM micrographs were taken by Yoshio Yamada, and light microscopic micrographs and illustrations are the work of Yozo Iwanami for this publication. The text was written by all three of those whose names appear as authors. We wish to express our gratitude to all those who helped us publish this volume, and give special thanks to Mr. Shuichi Muto, and Mr. Ippei Ohta of Kodansha Scientific.

Winter, 1987

THE AUTHORS

Contents

1. Genesis of Pollen

1.1 Flower—reproductive organ of plant

Plants, like animals, have male and female sexuality. Many plant species undergo sexual reproduction to producc seeds. The flower of plants is an organ for sexual reproduction. There are various types of flowers, but the shape of the flower is species specific and one of the best criteria for species classification and identification of relationships between species.

The standard structure of the flower of angiosperms consists of four parts: pistil, stamen, petal and calyx (representative types are the flowers of rape, cherry and camellia). These organs are called floral leaves in botany because they evolved from leaves in the process of angiosperm evolution. Each floral leaf is not necessarily distinct from each other in all species. There exist variations such as non-differentiation between petal and calyx in the lily, petalization of stamens as exemplified by the double flower of the camellia, and leaf-like stamen structure such as is found in the cherry blossom.

Fig. 1.1 SEM micrograph of pollen grains in the mature anther locule of Easter lily (*Lilium longiflorum*) at flowering stage. p: pollen grain. t: degraded tapetum layer. e: endothecium.

Among four floral leaves, the pistil and stamen in which female and male gametes are generated, are the most important parts for reproduction (Fig. 1.2). The pistil consists of three parts: stigma, style and ovary. The ovary contains ovule(s) in which the egg cell(s) is included, and the sperm cells (male reproductive cells) of plants are contained in the anthers of the stamen (Fig. 1.3).

Fig. 1.2 Reproductive organs of plant (1)—pistil (female organ) and stamen (male organ).
A: water lily (*Nymphaea tetragona*). B: cactus (*Cactacea* sp.). C: yucca (*Yucca recurvifolia*).

Fig. 1.3 Reproductive organs of plant (2).
A: broom (*Cytisus scoparius*). B: broad bean (*Vicia faba*). C: Indian pipe (*Monotropastrum globosum*). D: starwart (*Enigeron annus*).

The number of ovules in an ovary is relatively small (one in the cherry blossom, and a few in others), compared with number of the male gamete cells, *i.e.* pollen. The stigma at the tip of the pistil has branch- or hairlike structures and secretes fluid to receive pollen grains. On the other hand, the stamen produces abundant pollen to increase the probability of mating with the female gamete (Fig. 1.1). It is by means of these morphological structures of flowers that plants propagate to continue the species.

1.2 Pollen mother cell and microspore

Pollen mother cells (microsporocytes) are differentiated from sporogenous cells in the young male organ. These cells are poorly attached to each other by plasma connections. During growth, the pollen mother cells become spherical and then separate from each other. A thick callose layer is formed surrounding the pollen mother cells; this callose, β-1, 3- glucan, is an amorphous cell-wall substance. The pollen mother cells undergo meiosis to produce tetrads, which in the haploid phase are called microspores. The members of the tetrad become separated from one another by a callose wall which is continuous with the callose surrounding the entire tetrad. The prevailing arrangements of the members of the tetrads are tetrahedral or tetragonal depending on the orientation of the meiotic spindle axes and of the related cleavage planes. These processes are illustrated in Figs 1.4–1.6.

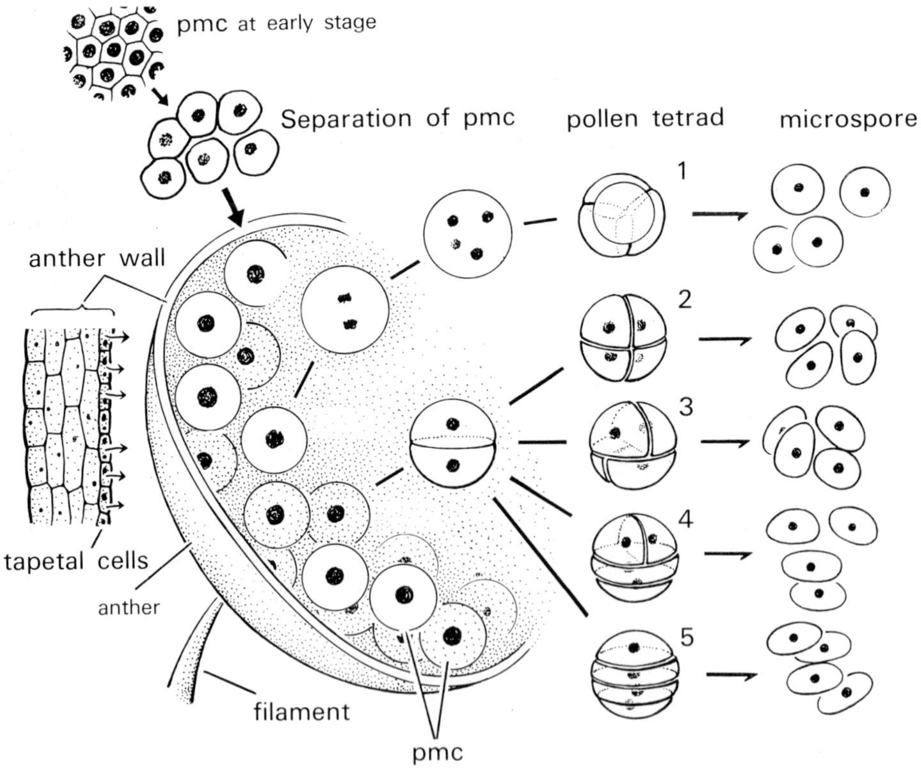

Fig. 1.4 Genesis of microspore.
pmc: pollen mother cell. 1–5: five orientations of pollen tetrad.

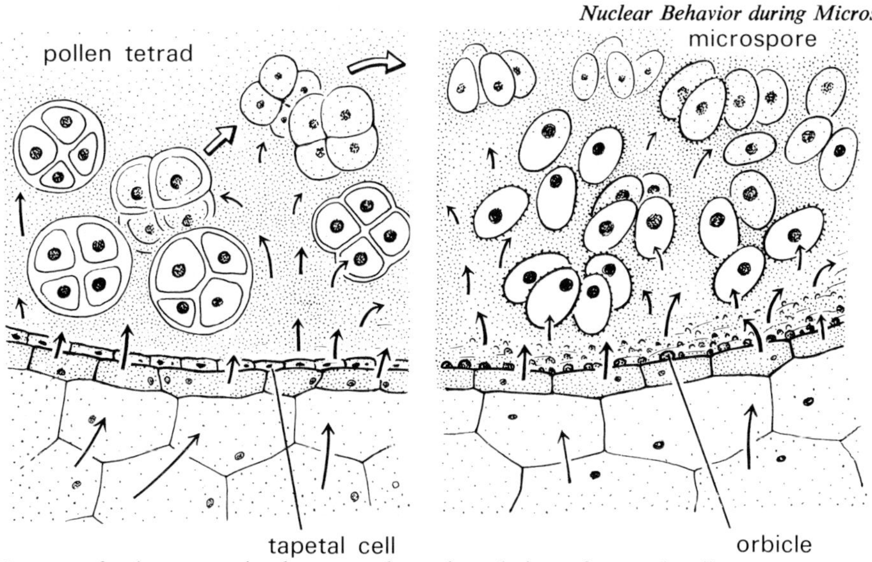

Fig. 1.5 Process of microspore development along degradation of tapetal cells.

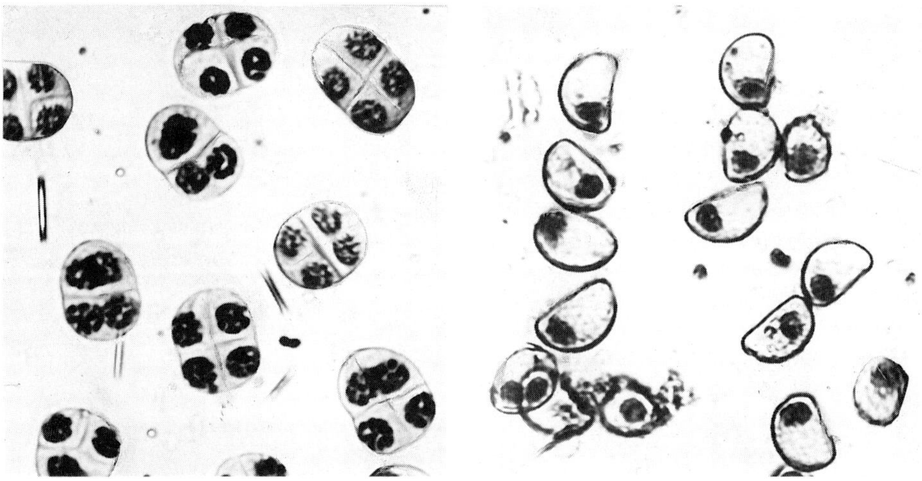
Fig. 1.6 Microspore development of lily.

The tetrad stage is important in the development of the pollen grain. The surface features of the mature grain are clearly related to the original orientation of the microspore within the tetrad. Within each microspore, peripheral elements of the endoplasmic reticulum are fixed which polymerize sporopollenin to form the sexine matrix on its surface. At the time the sexine structure is established, the callose disappears. Later, the sexine thickens without changing its structure. Each microspore in turn differentiates into a pollen grain.

1.3 Nuclear behavior during microsporogenesis

When the pollen mother cell undergoes meiosis, the nucleus changes to become chromosomes. Meiosis is the process of two successive nuclear divisions for one nuclear DNA replication (see Fig. 1.10). In *Tradescantia*, or *Lilium*, reductional division occurs at the first nuclear division, but, on the contrary, *Luzula* sp. undergoes reductional division at the second division. In any case, the endproducts of meiosis are cells in haploid (*n*) phase.

Four cells are produced around the four nuclei produced after meiosis to form a tetrad. These four cells then detach from each other to become microspores. The nucleus of a microspore cell undergoes further nuclear division (first pollen mitosis) to form a dinuclear cell. A cell membrane forms around one nucleus, and the cell continues the process of pollen maturation. A mature pollen grain therefore consists of two cells, one enveloped in the other. The larger cell is called a vegetative cell containing a vegetative nucleus, and the enclosed cell is a generative cell with a generative nucleus which will undergo further mitotic division (second pollen mitosis, or pollen tube mitosis). These processes are illustrated in Fig. 1.7 for a case of common spidewort (*Tradescantia virginiana*). Each step is correlated to bud length as shown in parentheses. Under the light microscope, a pollen grain of common spidewort appears to have two nuclei, one round and the other elongated. Electron microscopic observation reveals, however, that the smaller nucleus is surrounded by independent cytoplasm enclosed by a fully developed cell membrane.

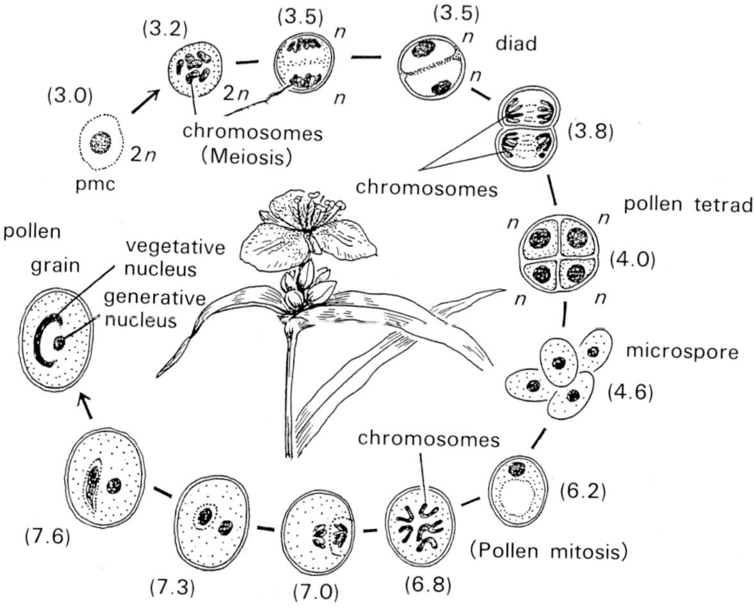

Fig. 1.7 Nuclear behavior during microsporogenesis and pollen maturation (dinuclear type).

1.4 Dinuclear pollen and trinuclear pollen

A microspore differentiates to become a cell with two nuclei, vegetative and generative, after a mitotic division (first pollen mitosis). The cell matures to become a pollen grain having two nuclei in some species; this is called dinuclear pollen. This type includes species of camellia (*Camellia japonica*), common spidewort (*Tradescantia reflexa*), and touch-me-not (*Impatiens balsamina*). On the other hand, pollen of rice (*Oryza sativa*), wheat (*Triticum aestivum*), corn (*Zea mays*), chrysanthemum (*Chrysanthemum* sp.), sunflower (*Helianthus* sp.), or rape (*Brasica* sp.) have three nuclei in a mature pollen grain, which is formed after a further division of the generative cell (second pollen mitosis). This type of pollen, called trinuclear pollen, has one vegetative nucleus and two sperm nuclei in each grain (Figs. 1.8, 1.9).

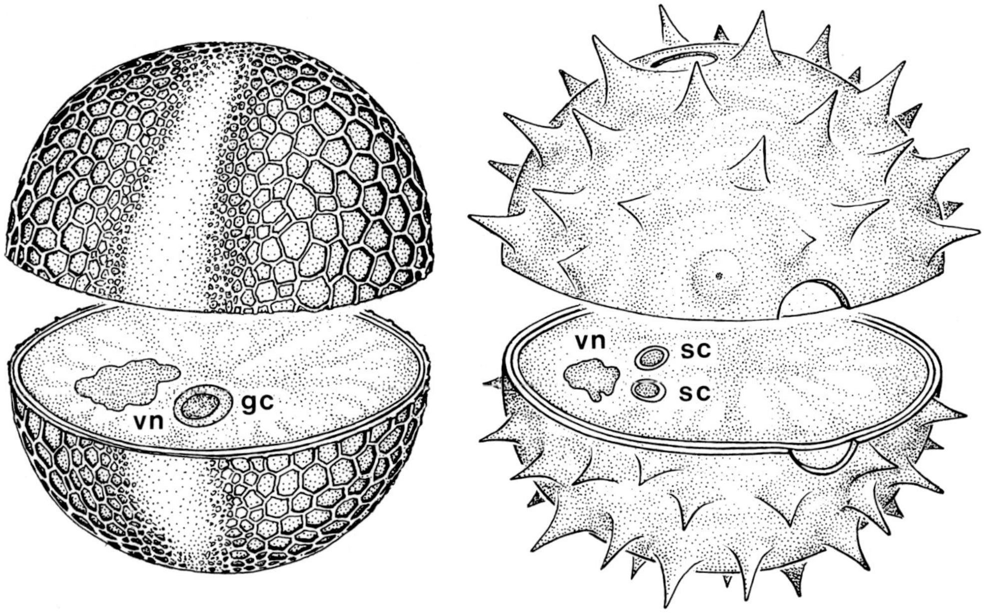

Fig. 1.8 Dinuclear pollen (*left*) and trinuclear pollen (*right*).
vn: vegetative nucleus. gc: generative cell. sc: sperm cells.

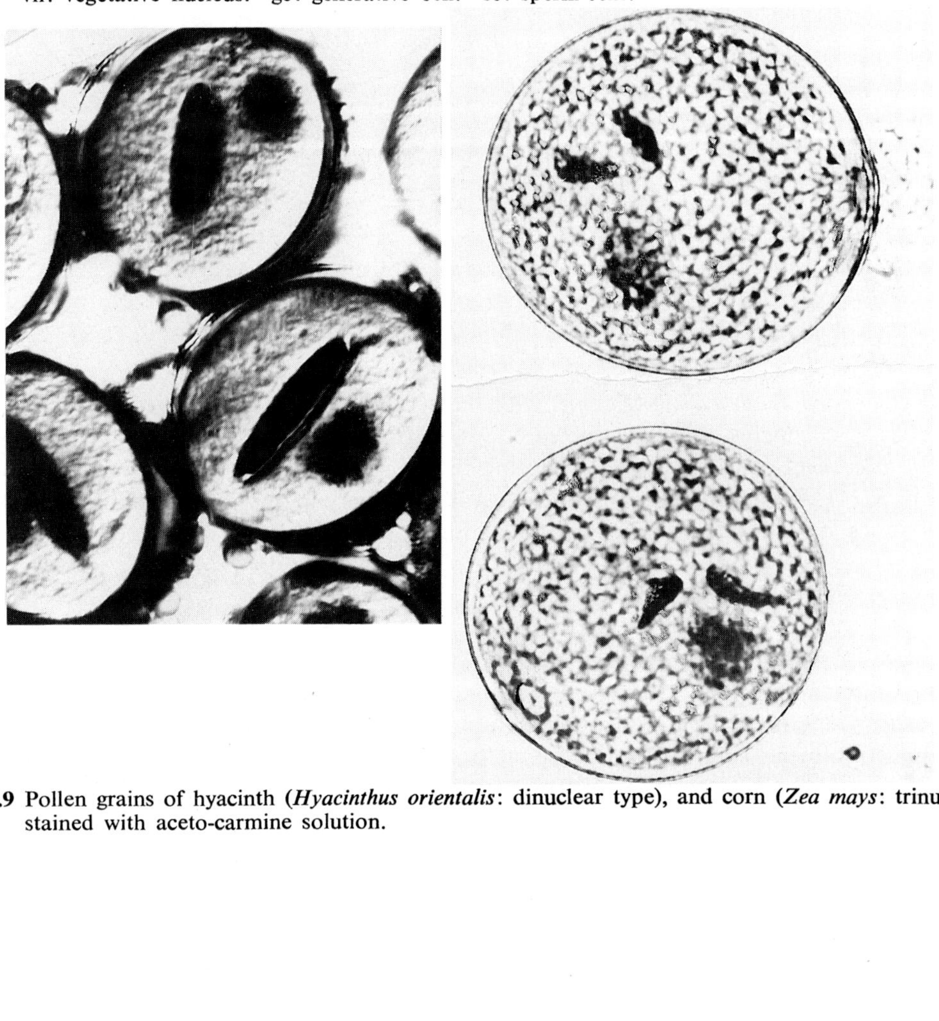

Fig. 1.9 Pollen grains of hyacinth (*Hyacinthus orientalis*: dinuclear type), and corn (*Zea mays*: trinuclear type) stained with aceto-carmine solution.

Figure 1.10 is a comparison of di- and trinuclear pollen types. In both case, mature pollen grains are released from anthers during the M-phase of the cell division cycle (mitotic nuclear division stage) after passing through the G_1-phase (preparatory stage for DNA synthesis), S-phase (DNA synthesis stage) and G_2-phase (preparatory stage for nuclear division). It should be noted that the trinuclear type releases pollen grains during the late M-phase with two distinctive sperm nuclei, while the dinuclear pollen type produces an undivided generative nucleus containing twice the DNA of the sperm nucleus. Generative nucleus division takes place in the pollen tube after the pollen grain has germinated on the stigma.

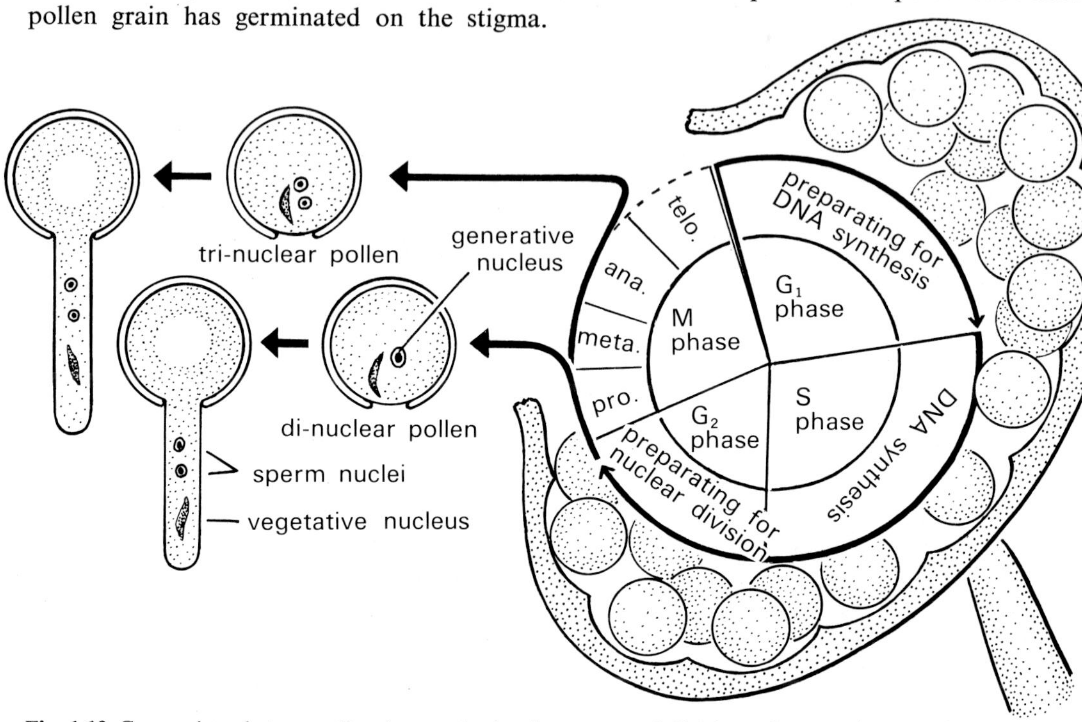

Fig. 1.10 Comparison between dinuclear and trinuclear types of division of generative nuclei.

1.5 Mononuclear pollen

When branches of *Tradescantia* (*Lilium*) with buds of varying sizes are incubated in a bottle containing 0.1% colchicine solution, nuclear division of the microspore (first pollen mitosis) is inhibited by the effect of the chemical, resulting in the production of pollen grains having one nucleus (Fig. 1.14). These artifical mononuclear pollen grains can normally germinate, elongating the pollen tubes on the stigma or artificial medium.

1.6 Pollen grain production

The amount of pollen grain produced by a flower varies by species as well as plant age and climatic environment. Table 1.1 shows the average amount of pollen grain produced by flowers of five species under natural conditions. For example, a flower of camellia releases 1,800,000 pollen grains from 180 stamens, each containing about 10,000 pollen grains. Considering that a camellia tree bears thousands of flowers, the number of pollen grains which are released into the air in the flowering season is beyond imagining.

Generally speaking, the anemophilous flowers (pollinated by wind) produce much more pollen than the entomophilous flowers (pollinated by insects). As an example of the anemophilous type, a floret of pine produces 1,480,000 pollen grains and an influorescence consists of an average of fifteen florets. Therefore, one pine tree bearing 200 influorescences releases 1,480,000 × 15 × 200, or 4,440,000,000 grains of pollen.

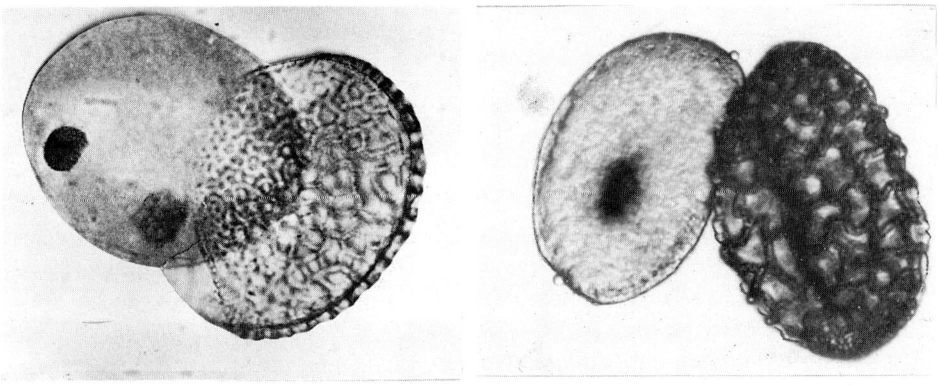

Fig. 1.11 Normal pollen grains with two nuclei (*left*), and artifically induced mononuclear pollen with colchicine treatment (*right*) of lily.

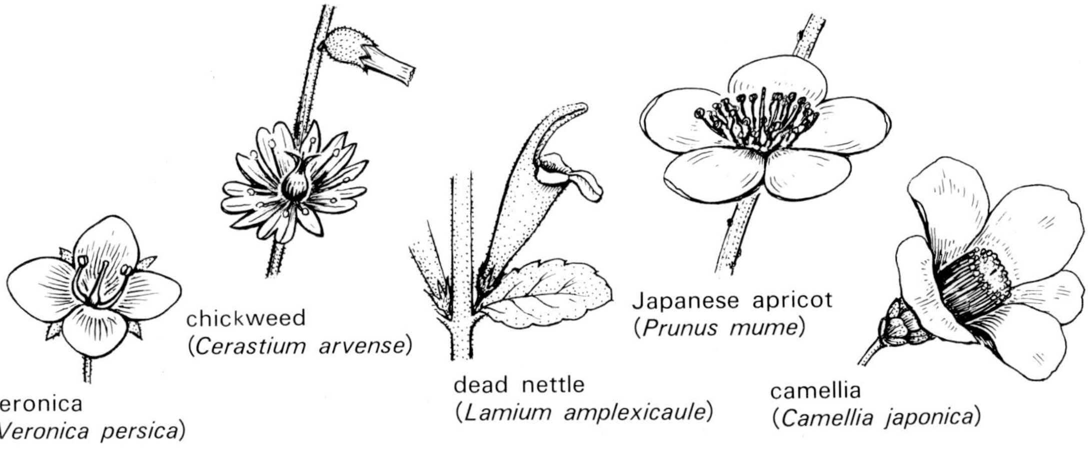

Fig. 1.12 Variation of flower morphology: types and numbers of stamens producing pollen grains.

TABLE–1. Average Amount of Pollen Grains by Flowers of Five Species

	Veronica	Chichweed	Dead nettle	Japanese apricot	Camellia
No. of pollen grains in an anther	2987	257	847	1290	10021
No. of stamens	2	10	4	51	180
No. of pollen grains in a flower	5974	2570	3388	65790	1803780

2. Pollen Morphology of Flowering Plants

2.1 Size of pollen grain

The seize of one pollen grain is generally less than 0.1mm in diameter, but there is wide variation among plant species (Fig. 2.2). One of the largest pollen grains is that of *Zostera marina*, having a long thread-like structure 2–3 mm in length. Morning glory (*Pharbitis nil*), bottle ground (*Lagenaria leucantha*), four-o'clock (*Mirabilis* sp.), ginger (*Zingiber officinale*), hollyhock (*Alcea rosea*), cotton rose (*Hibiscus mutabilis*), and pumpkin (*Cucurbita* sp.) produce relatively large pollen grains, while the pollen grains of forget-me-not (*Myosotis alpendtris*), iwatabako (*Conandron* sp.), and bluebells (*Merteusia asiatica*) are less than 10 μm in diameter. Standard size grains are found in lily-of-the-valley (*Convallaria* sp.), violet (*Viola* sp.), gentian (*Gentiana* sp.), cosmos (*Cosmos* sp.), plantain (*Plantago* sp.), ginkgo (*Ginkgo biloba*), Japanese red cedar (*Cryptomeria japonica*), birch (*Betula tauschii*), cherry (*Prunus yedonensis*), Japanese apricot (*Prunus mume*) with diameters of about 50 μm. In general, pollen of anemophilous (wind-pollinated) flowers are smaller than those of entomophilous (insect-pollinated) flowers.

Four microspores from a pollen mother cell separate from each other when they develop to become pollen grains in the pollen sac. However, in some species such as azalea (*Azalea* sp.), reed-mace (*Typha* sp.) and venus fly-trap (*Dionaea muscipula*), the tetrads do not separate but remain attached to each other even when they mature to form a pollen mass of four or sixteen grains (Fig. 2.1).

Pollen grains of some species have cord-shaped structures called viscin threads on the surface, which function as holding devices for grain attachment or for hanging on insect legs (Fig. 2.1).

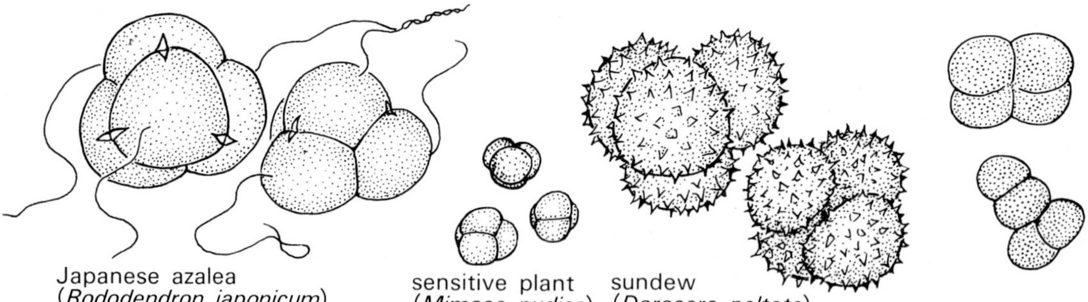

Fig. 2.1 Tetrad pollen grains.

Fig. 2.2—*(Continued from p. 11)*
rose (*Hibiscus mutabilis*) 16: Japanese red cedar (*Cryptomeria japonica*). 17: Japanese hop (*Humulus japonica*). 18: nettle tree (*Celtis sinensis*). 19: spruce (*Picea jezoensis*). 20: umbrella pine (*Sciadopitys verticillata*). 21: birch (*Betula platyphylla*). 22: chestnut (*Castanea crenata*). 23: dahlia (*Dahlia pinnata*). 24: oak (*Quercus acutissima*). 25: gogyou (*Gnaphalium multiceps*). 26: siberian yellow (*Achillea sibirica*). 27: iwatabako (*Conandron ramondioides*). 28: burrebur (*Petasites japonicus*). 29: sasa (*Sasa veitchii*). 30: bell-ware (*Zostera marina*). 31: bog-violet (*Pinguicula vulgaris*). 32: foxtail (*Alopecurus aequalis*). 33: turf (*Zoysia japonica*). ·34: rice (*Oryza sativa*). 35: indian corn (*Zea Mays*). 36: wheat (*Triticum aestivum*).

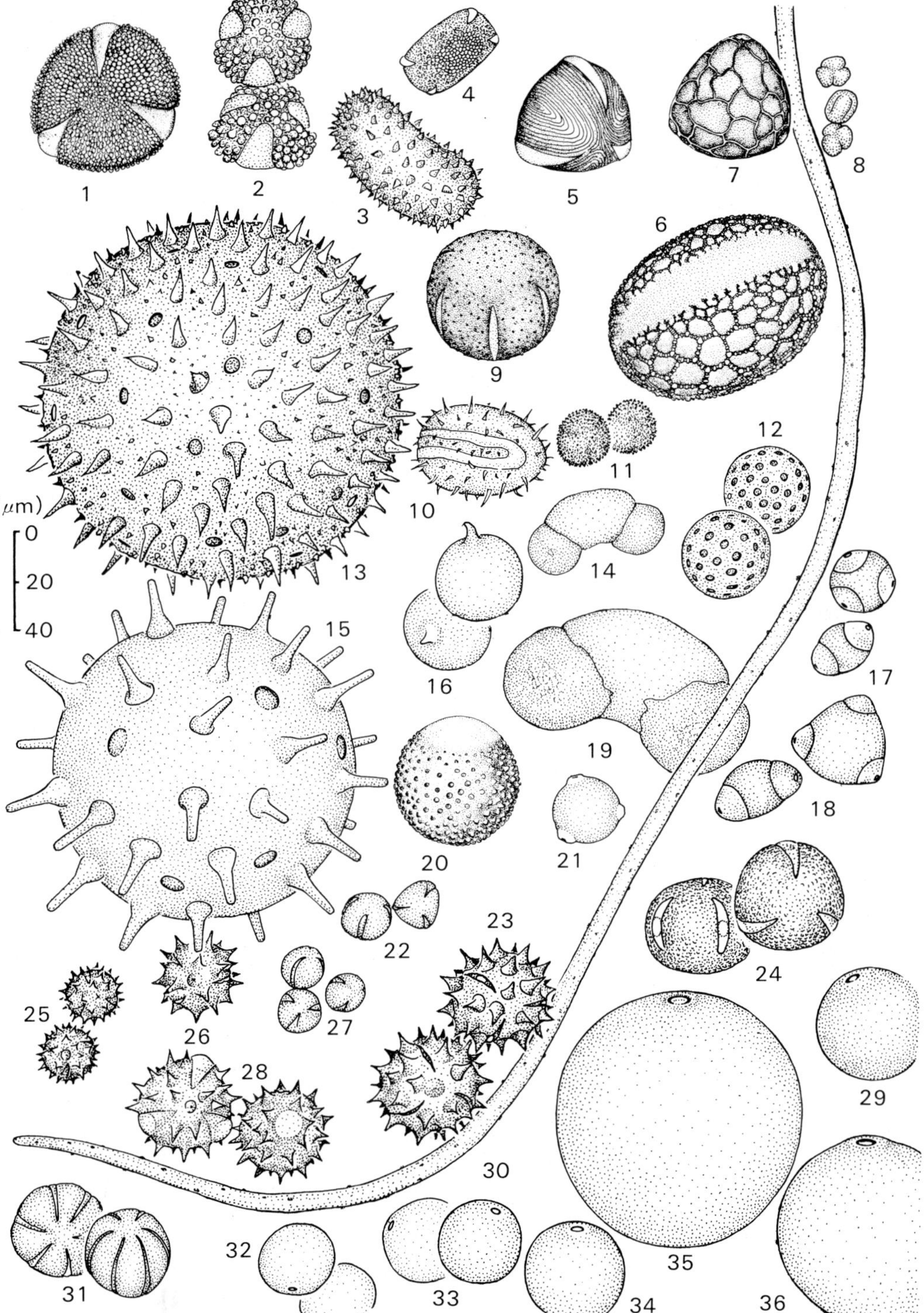

Fig. 2.2 Various types of pollen grain.
1: aucuba (*Aucuba japonica*). 2: holly (*Ilex integra*). 3. day flower (*Commelina communis*). 4. touch-me-not (*Impatiens balsamina*). 5: cherry (*Prunus yamasakura*). 6: lily (*Lilium longiflorum*). 7. azuki-bean (*Phaseolus angul'aris*). 8: iwatabaco (*Conandron ramondioides*). 9: baloon flower (*Platycodon grandiflorum*). 10: brandy bottle (*Nuphar japonicum*). 11: ragweed (*Ambrosia artemisiifolia*), 12: spinach (*Spinacia oleracea*). 13: hollyhock (*Althea rosea*) 14: red pine (*Pinus densiflora*). 15: cotton

2.2 Saccate pollen grain

Most species of the pine family (*Pinaceae*) produce pollen grains with a pair of air sacs (sacci). These are called saccate pollen grain and are produced by, for example, red pine (*Pinus densiflora*), black pine (*P. Thunbergii*), white pine (*P. pentaphylla*), dwarf stone pine (*P. pumila*), veith (*Abies ventchii*), fir (*A. firma*), deodar cedar (*Cedrus deodara*), spruce (*Picea excela*) and *Podocarpus* sp. (Fig. 2.3). On the other hand, larch (*Larix leptolepis*) and hemlock fir (*Tsuga ieboldii*) do not have air sacs on their pollen.

Taking advantage of air resistance, these saccate pollen can migrate far through the air. The shape of the air sacs varies by species, and the types can be identified by comparing the relative distance between the grain body and the sacs. When they are migrating through the air, the grains shrink due to dehydration and become fully expanded when they absorb water in the stigma (Figs. 2.4, 2.5). The pollen tube will germinate from the narrow sulcus (furrow) of the grain body between the sacs.

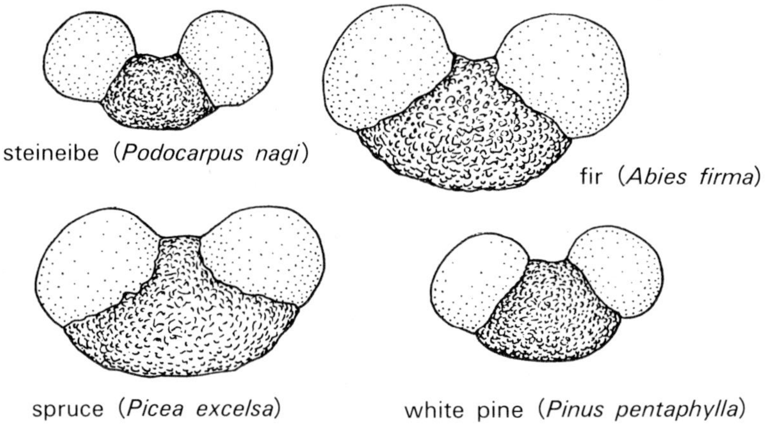

Fig. 2.3 Saccate pollen (pollen grains with air sacs).

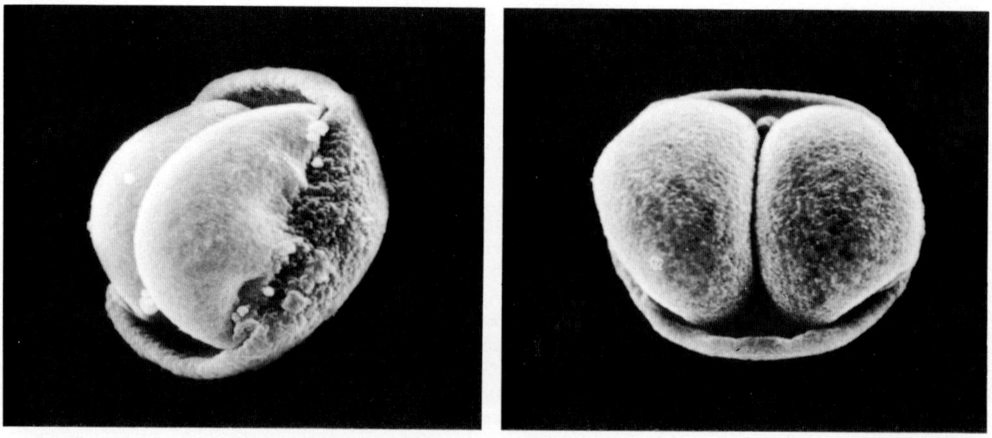

Fig. 2.4 Pollen grains of black pine (*Pinus Thunbergii*). Because of the air sacs, pollen grain appears different from different views. ×950.

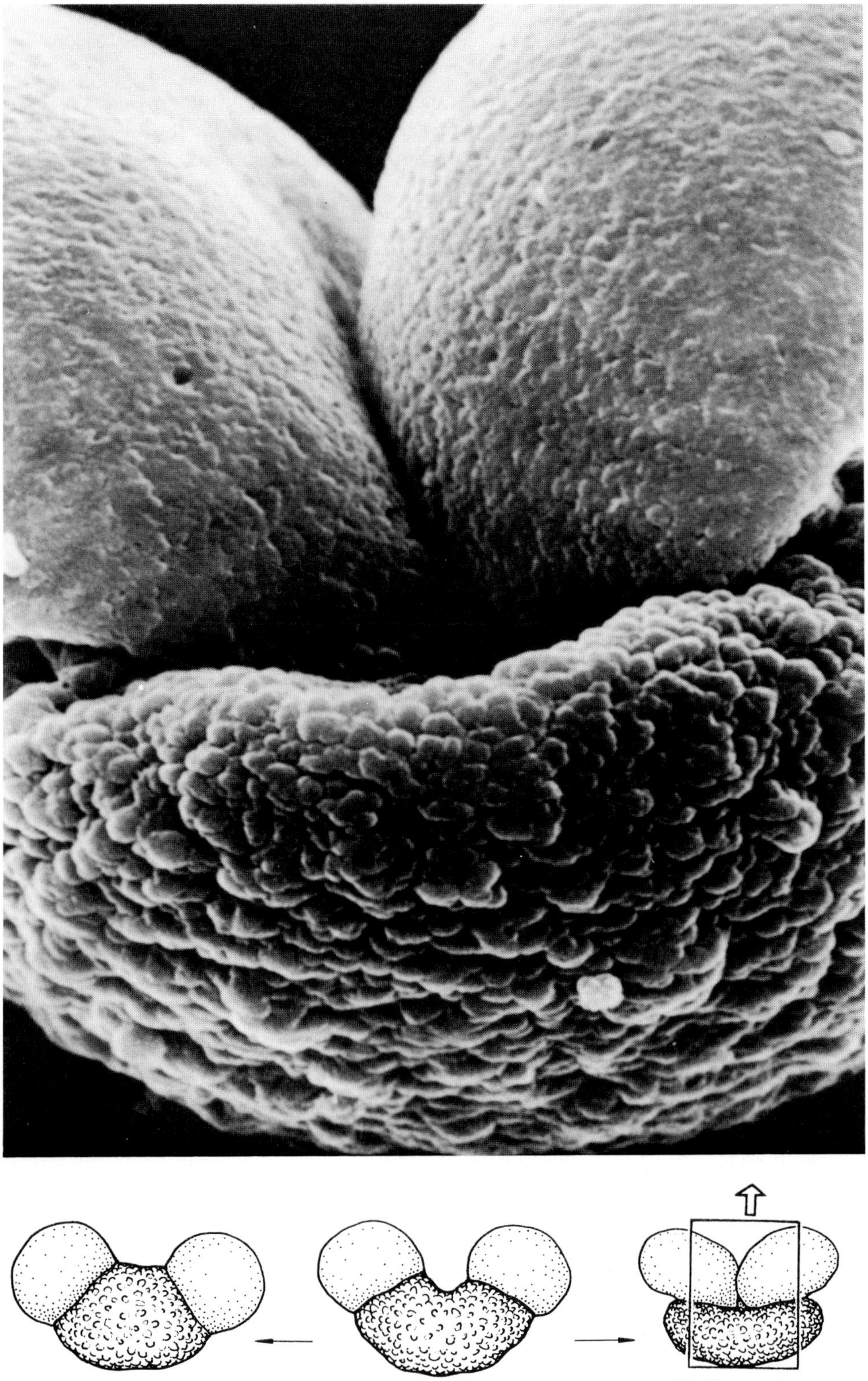

Fig. 2.5 Surface morphology of black pine pollen (*Pinus Thunbergii*) and changes through hydration and dehydration. The pollen grain is seen magnified 5,000 times.

2.3 Pollen wall stratification

Pollen grains of most angiosperms resemble other plant cells, containing cytoplasm surrounded by a thin wall of cellulose membrane called the intine. Immediately outside the intine is another wall known as the exine. The typical exine of angiosperm pollen grains consists of two layers: an inner, basal layer, or nexine ("non-sculptured part"), and an outer layer, or sexine ("sculptured part"). Figure 2.6 shows the stratification of the wall of the angiosperm pollen grain.

The major chemical component of the exine is sporopollenin, polymers of carotenoids and carotenoid esters. Sporopollenin is broken down at 500° C with sooty flame, but the unoxidized sporopollenins are remarkably stable and resistant against various chemicals such as aqua regia and sulfuric acid.

The apertural sites are parts of the wall where the sporopollenin is deposited in granules or in thin sheets and the underlying intine is exposed.

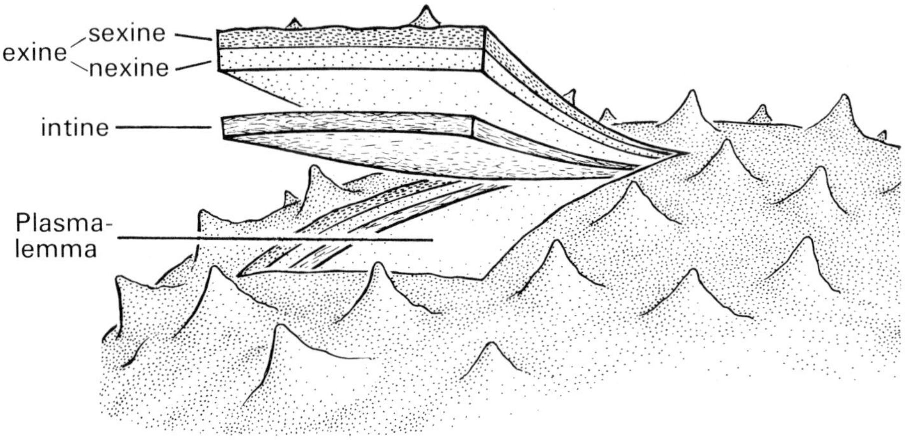

Fig. 2.6 Generalized structure of pollen wall. Sexine has species specific sculpture, and inside the intine, the cytoplasm and nuclei are enveloped by the plasmalemma.

2.4 Ornamentation of the exine

The outer surface of the exine is marked with various kinds of sculpture and ornamentation. The sculpture of the pollen grain is a fairly constant characteristic which is, in most cases, an excellent means of identification of species. Although the patterns of these sculptures are difficult to describe in exact terms, and consequently somewhat unsatisfactory for analytical description, the major types of exine sculpturing are represented as follows: (1) psilate or smooth and lack sculpturing; (2) foveolate or pitted; (3) fossulate or grooved; (4) scabrate with very fine projections, usually of less than 0.5 μm; (5) verrucate or warty; (6) baculate with rodlike sculptured elements; (7) pilate (clavate) with rodlike sculptured elements with swollen heads, *i.e.* with pila (clavae); (8) gemmate with sessile pila; (9) echinate or spiny; (10) rugulate with elongate sculptured elements irregularly scattered over the pollen surface; (11) striate with elongated and more or less parallel sculptured elements distributed tangentially on the pollen surface; and (12) reticulate with sculptured elements forming an open network or reticulum on the entire pollen surface. Typical types of exine structure are illustrated in Figs. 2.7 and 2.8.

With regard to the surface of the exine, elaborate pollen sculpture seems correlated with entomophily, while psilate pollen grains are largely characteristic of anemophilous plants.

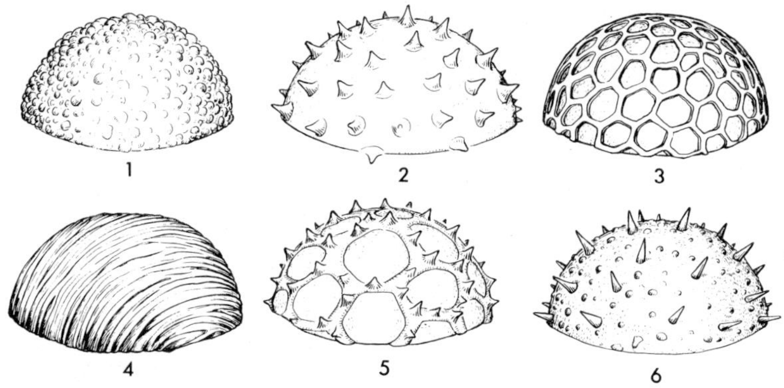

Fig. 2.7 Six typical patterns of exine structure.
1: knobby structure. 2: needle structure. 3: net structure. 4: finger print structure. 5: combination needle and net structure. 6: combination knobby and needle structure.

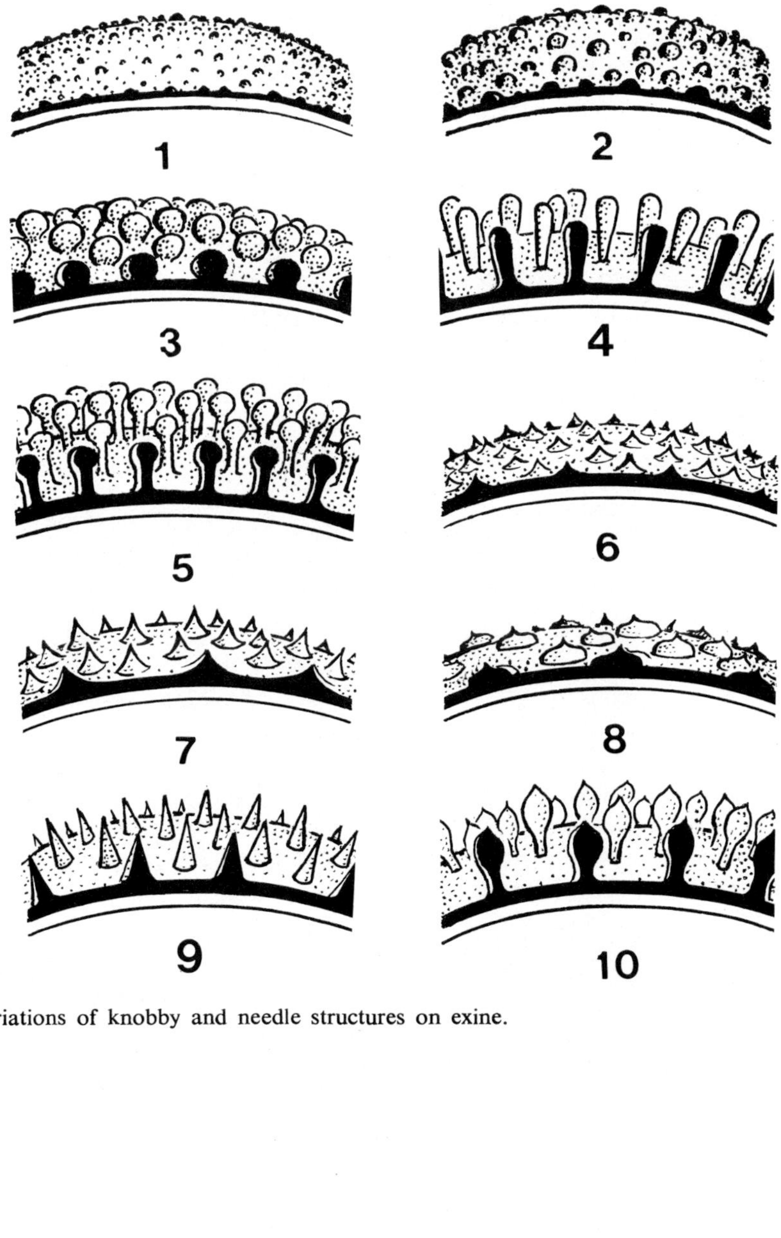

Fig. 2.8 Variations of knobby and needle structures on exine.

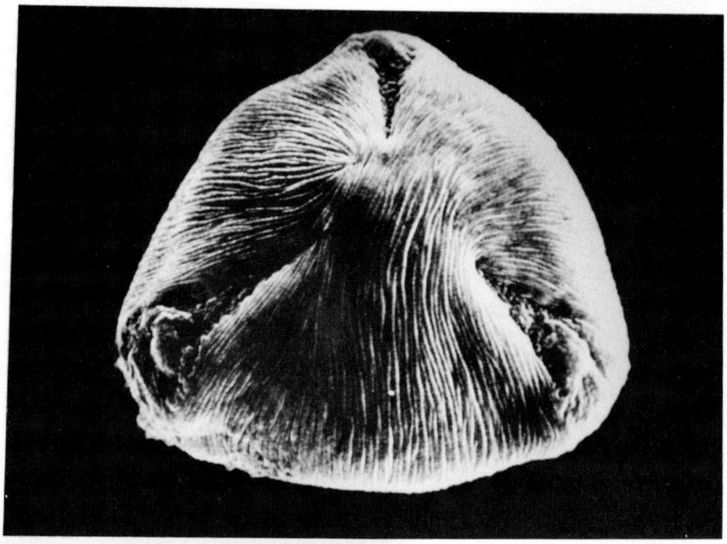

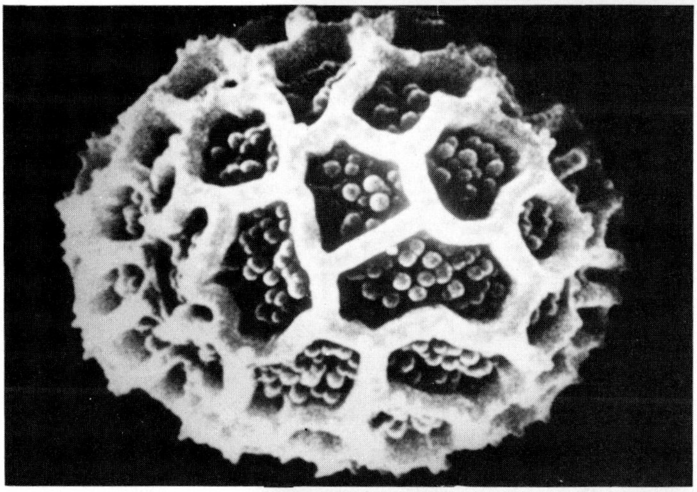

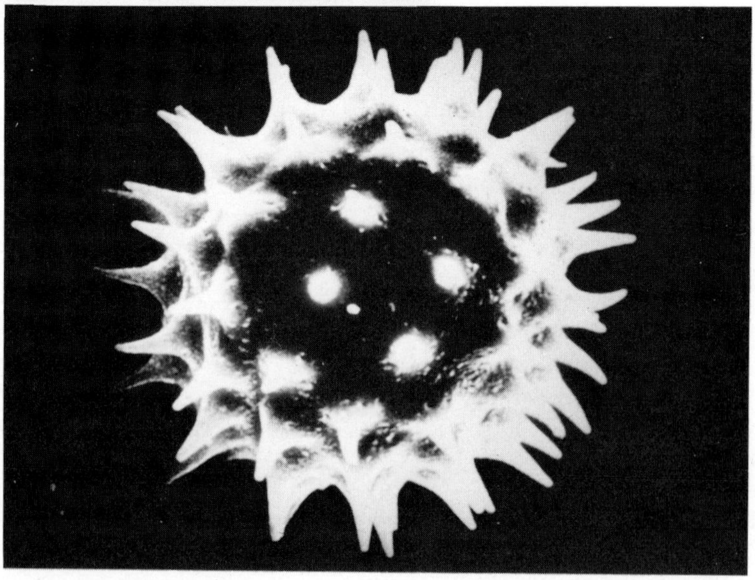

2.5 Apertures

Most pollen grain possess apertures (pores) in thin areas of the exine through which pollen tubes germinate. Two basic types of pollen grain are the porate and colpate (sulcate). Porate pollen grains, such as those of *Hibiscus*, have apertures in the exine (Fig. 2.38), while pollen grains of other species have long furrows called colpae (sulcae). When the pollen grain germinates, the pollen tube emerges through one of the apertures. The apertures may also permit volume changes caused by fluctuations in humidity.

The difference between furrows and pores seems to be purely morphological: furrows are elongated and boat-shaped and the ends are more or less sharp (Fig. 2.10). Pores are generally iso-diametric, and if they are elongated, the ends are rounded. Phylogenetically, furrows are apparently the primitive form; pores must have developed later by the contraction of furrows. Transitional forms occur. Generally, the arrangement of normal furrows and pores varies between species.

Pollen grains that have one long furrow are called monocolpate (monosulcate) (Fig.2.101). The monocolpate type is common in monocotyledons such as Easter lily (*Lilium longiflorum*) as well as in many primitive dicotyledons in *Ranales*. A large number of dicotyledons including Japanese apricat (*Prunus mume*) have pollen grains with a tricolpate structure (Fig.2.9). Tricolpate pollen grains have three long furrows that are equidistant from each other.

In another type of pollen grain, the central or median part of the aperture membrane shows distinct granules. These heavily covered apertures are called opercula and are distinct from the rest of the exine surrounded by a very thin membrane (Figs. 2.11, 2.12 and 2.13). In many grasses, they are easily lost when the grain expands with moisture. Beautiful opercula are found in the furrows of many *Compositae*.

Fig. 2.9 SEM micrographs of three entire pollen grains showing exine ornamentation.
Top: Japanese apricot (*Prunus mume*); Finely striate, polar view–angular to obtusely triangular, with convex side, the apertures situated in the middle of the side. ×1800. Middle: Bougainvillea (*Bougainvillea spectabilis*); Reticulum, lumina with numerous granules. ×2900. Bottom: Cuckold (*Bidens pilosa*); Distinct echinate, prominent spines. ×2000.

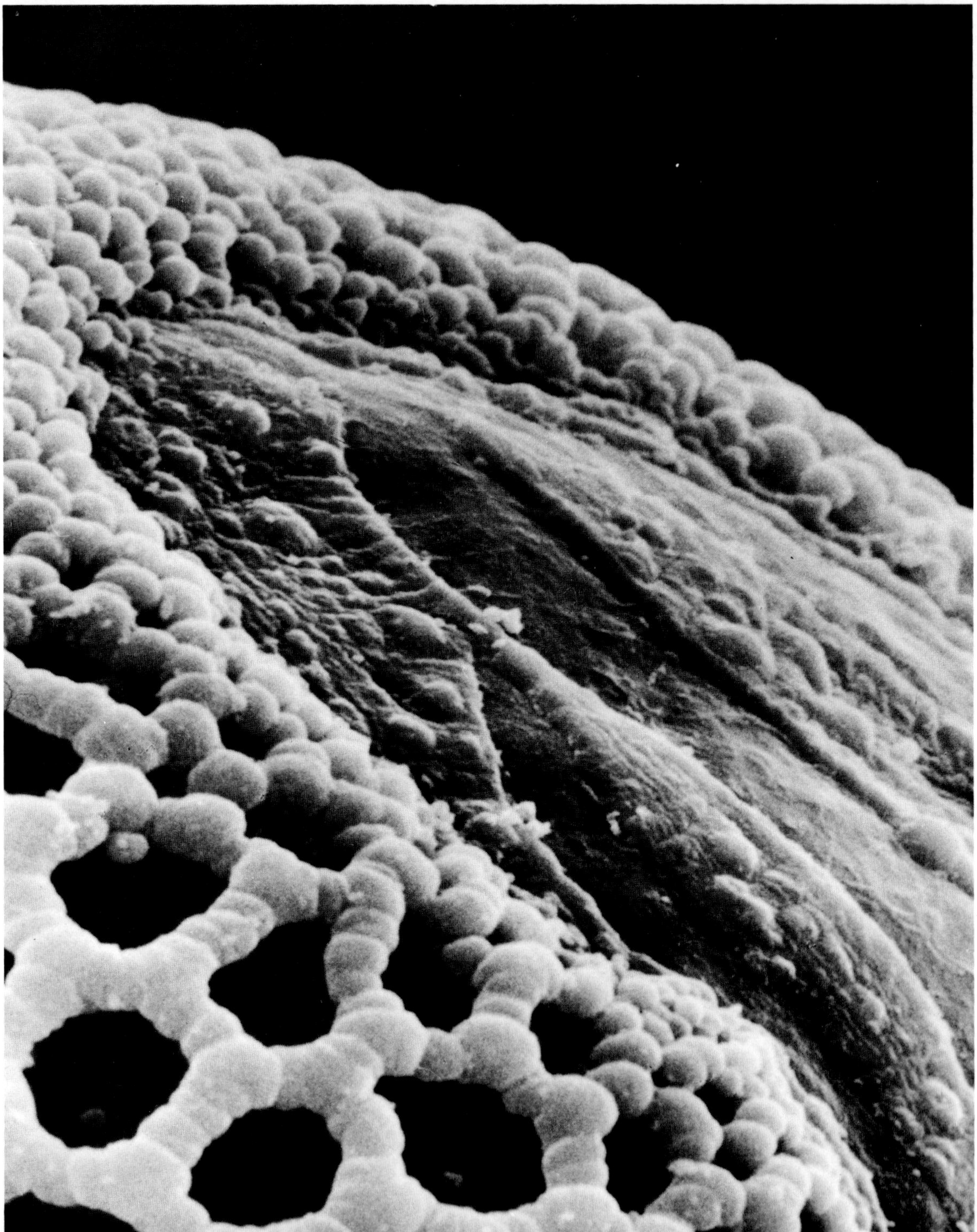

Fig. 2.10 Aperture furrow of Easter lily (*Lilium longiflorum*) pollen (×6700).

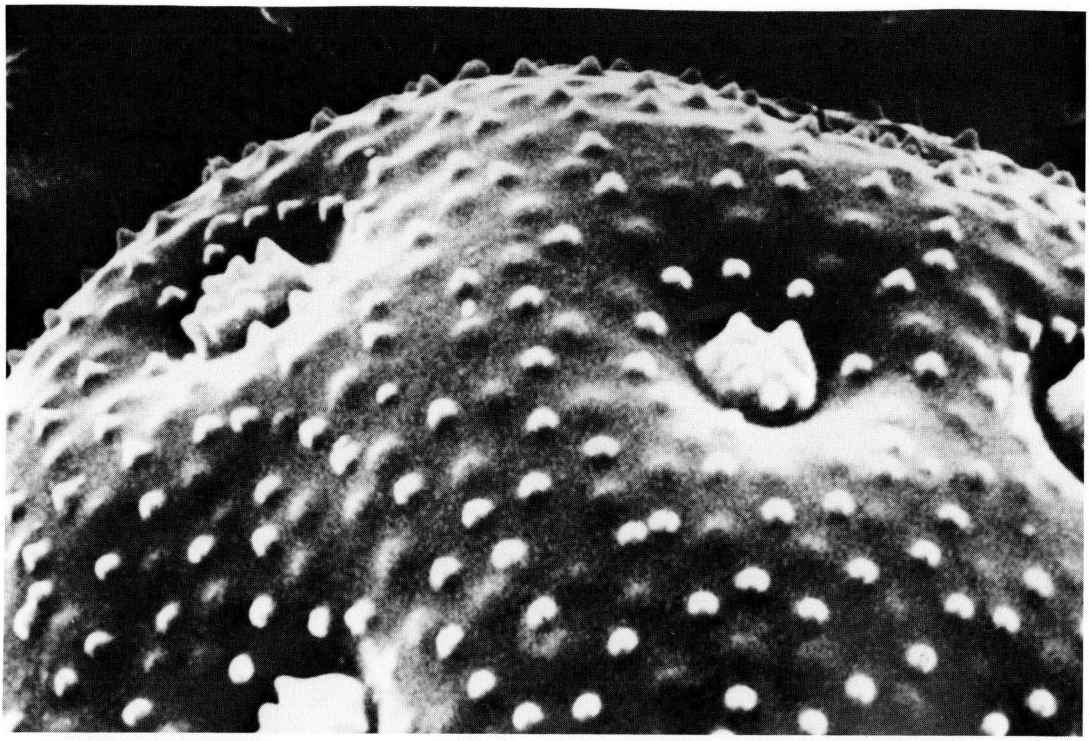

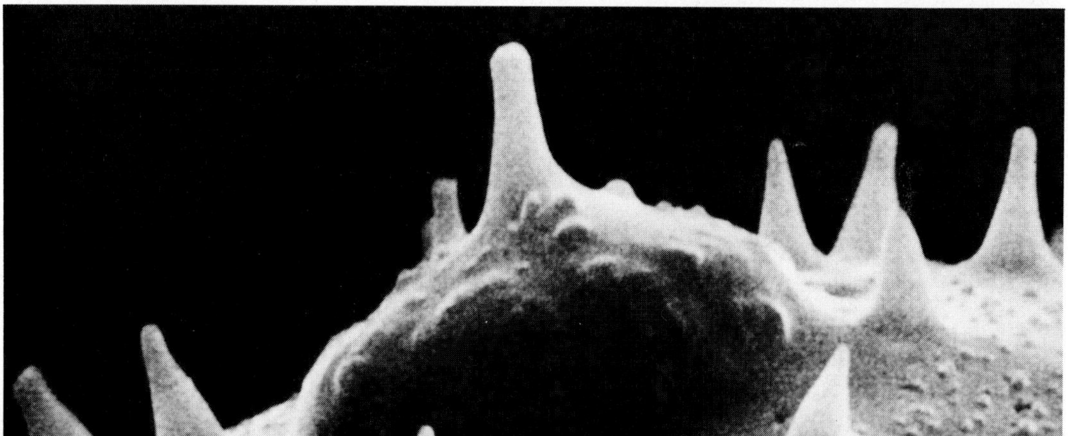

Fig. 2.11 Top: Pores (aperture) with plug-like operculum of garden pink (*Dianthus plumarious*) pollen. ×6500. Bottom: Pore with well-developed operculum of pumpkin (*Cucurbita moschata*) pollen. ×4500.

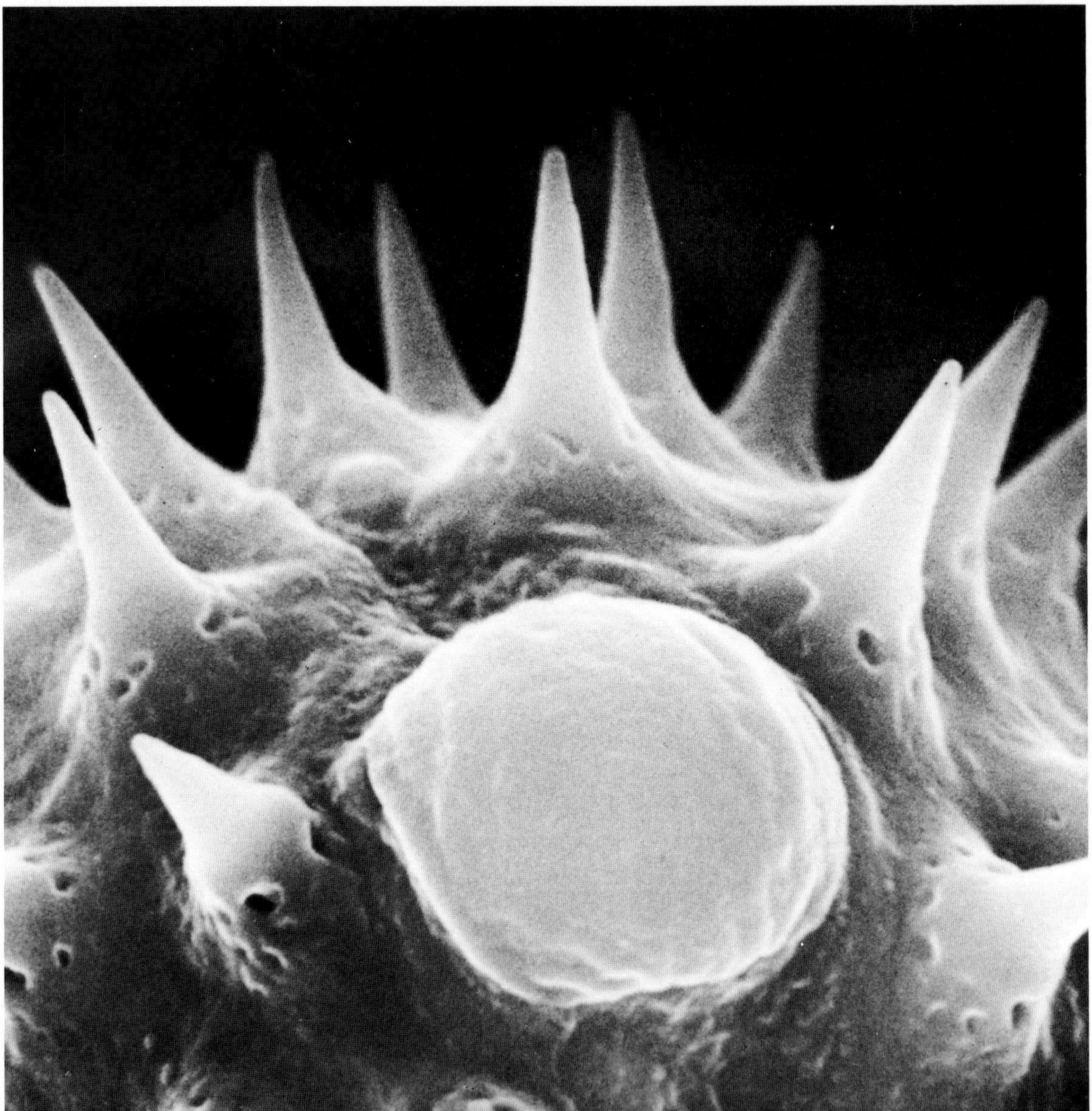

Fig. 2.12 Dome-like operculum (aperture cap) on sunflower (*Helianthus annuus*) pollen (×8000).

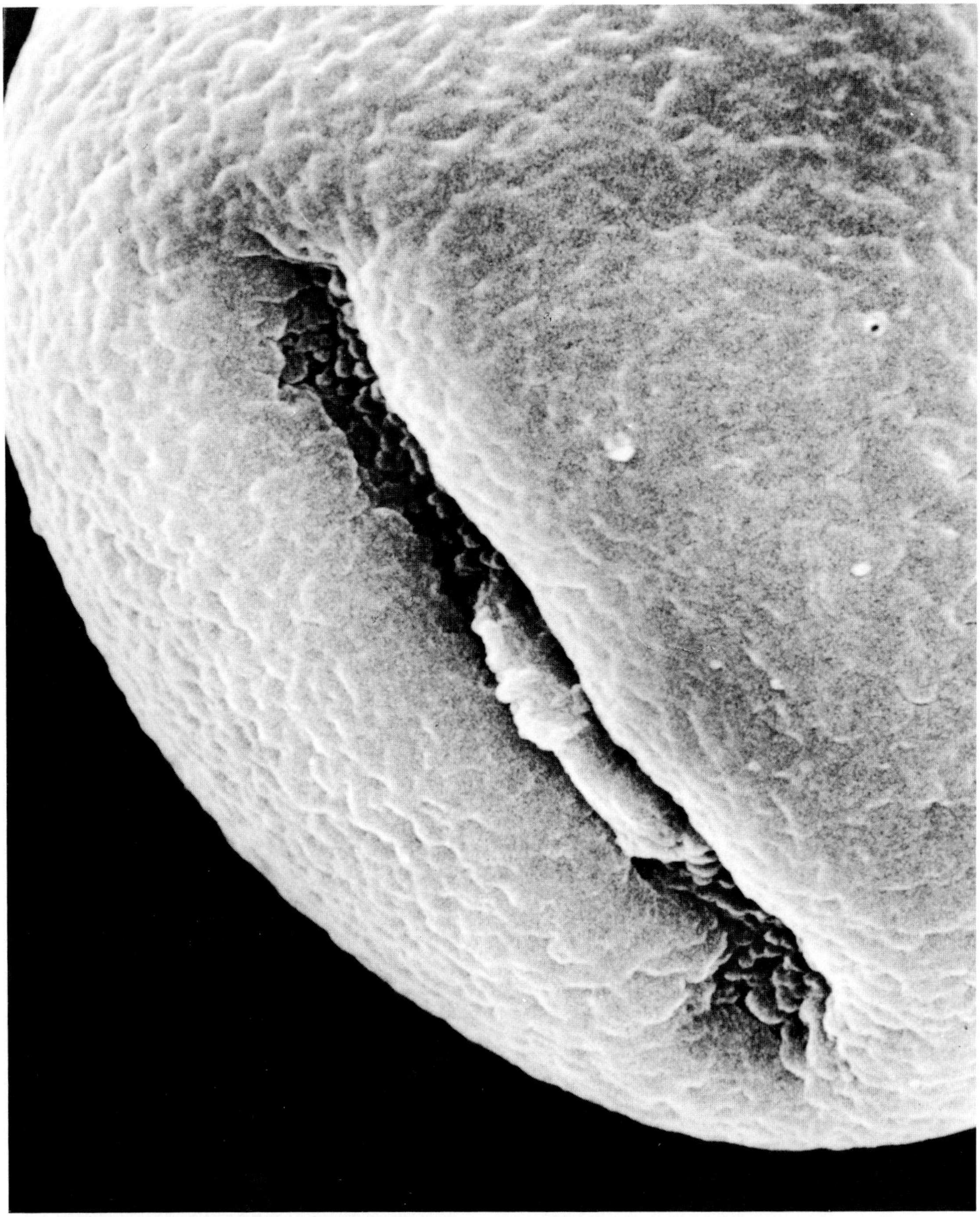

Fig. 2.13 Pollen surface with corpuscular projections of glorious bells (*Enkianthus perulatus*) ($\times$10000). A furrow aperture with operculum is seen in the central portion.

2.6 Contribution of the tapetum to the developing pollen grain

2.6.1 Tapetum

In the early stages of the development of pollen grains, sporogenous tissue is enclosed by a nutritive cellular layer termed the tapetum. Two main types of tapetum are recognized depending upon the development associated with sporogenesis; secretory and amoeboid. In the secretory (or glandular) type of tapetum (Fig. 2.14), the tapetal cells maintain a continuous position but become disorganized, eventually undergoing autolysis. In the amoeboid (or plasmodial) tapetum (Fig.2.15), tapetal cell walls break down at an early stage, forming a plasmodium which is associated with the developing pollen grains. Secretory tapeta are recorded more frequently in angiosperm families than amoeboid tapeta.

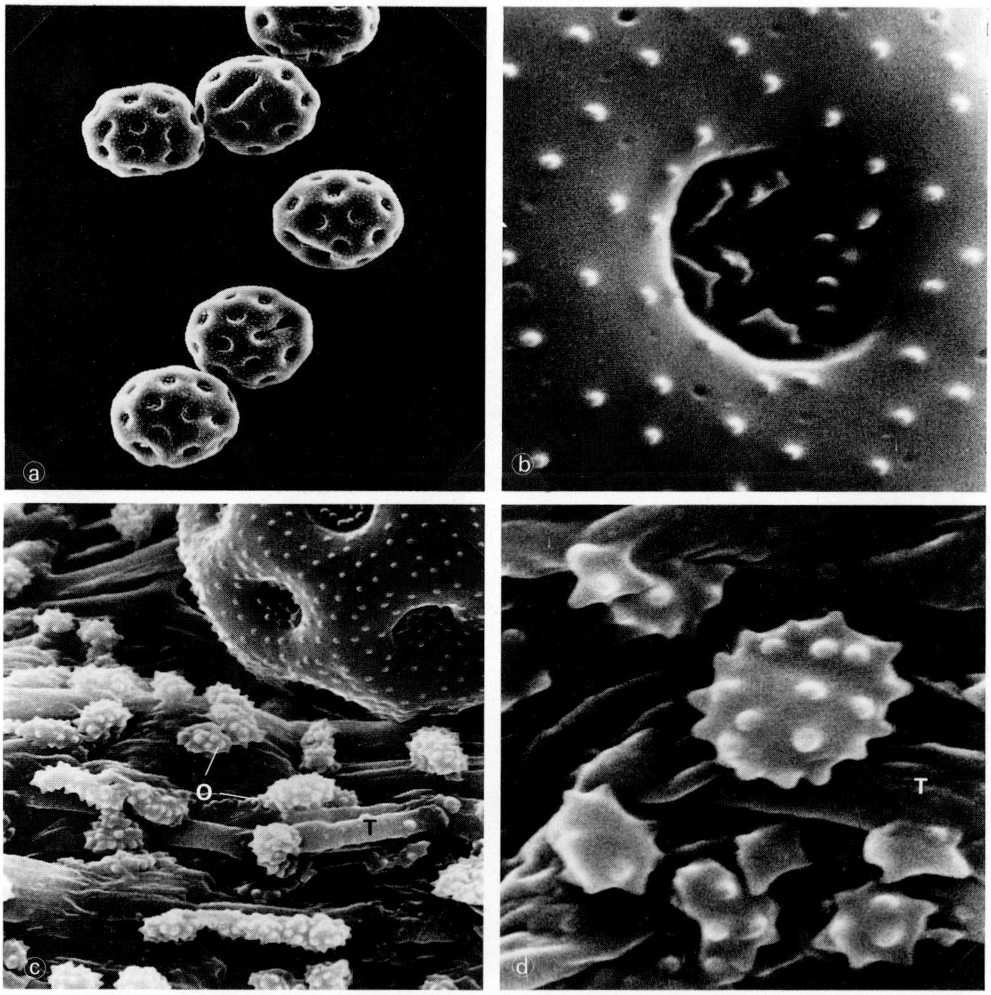

Fig. 2.14 SEM micrographs of secretory tapetum of catch-fly (*Silene armeria*).
A: Group of pollen grains. ×540. B: SEM micrograph of mature pollen grain, showing surface topography and germination pore (aperture). Ornamentation of exine showing blunt spinules in profile. ×7200. C: Face view of tapetal membranes, oriented towards anther locules, covered with orbicles. ×2160. D: Enlargement of a part of the same tapetum as in Fig. C. Each orbicule is somewhat irregular in size and shape. Note that the pattern of the surface of orbicules is identical to that of the exine surface. ×7200. T: tapetum, O: orbicule.

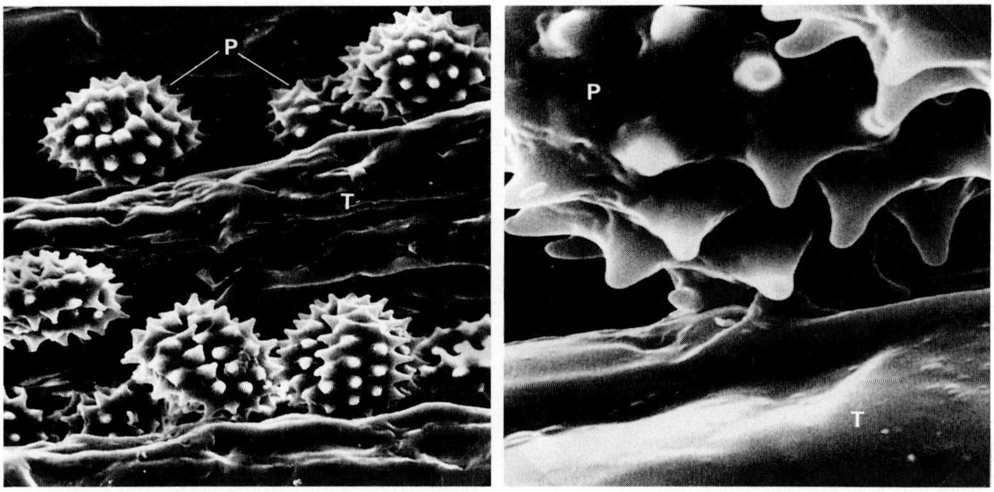

Fig. 2.15 SEM micrographs of amoeboid tapetum of thoroughwort (*Eupatorium japonica*). Inner tangential surface of tapetum after anther dehiscence (*left*). ×2160. In this species, tapetum is completely plasmodial. Pollen grain is closely attached to the wall of tapetum (*right*). ×7200. T: tapetum, P: pollen grain.

The tapetal plasmalemma breaks down when the microspores are vacuolate after having used up most of their cytoplasmic reserves. After the tapetal contents flow out into the locule, there is a dramatic increase in cytoplasmic density in the microspores. Vacuoles disappear, and considerable numbers of starch and lipid granules are seen. This may be indirect evidence that the major function of the tapetum is to provide food precursors to the microspores and young pollen grains at critical stages when they have almost exhausted their own reserves in producing an elaborate spore wall.

One of the distinguishing features of the secretory tapetum of many angiosperms is the production of particles on the inner locular face at approximately the tetrad stage of microsporogenesis. These particles have been termed orbicules or Ubisch bodies. Orbicules are spheroidal structures that are generally only a few micrometers in diameter; however, they may occasionally appear in large aggregates during the later stages of microsporogenesis. When mature, the wall of the orbicule appears uniformly thick and occasionally exhibits the same ornamentation as the pollen grain exine (Fig. 2.14). On the basis of chemical properties and resistance to acetolysis, orbicules appear to be formed through sporopollenin accretion. The biological function of the orbicules is uncertain, but they may play some kind of role in the dispersal of pollen.

2.6.2 Tryphine

During the growth of microspores into mature pollen grains, structural changes also occur in the cells of the tapetum. With the disruption of the tapetum during the later phase of pollen maturation, tapetal residues called tryphine pass into the anther cavity. Lipidophillic material is a major component of tryphine. These coalesced bodies may give rise to the sticky material that covers the mature pollen grains of most insect-pollinated plants. In *Liliaceae*, where the exine is reticulate, deposits of tryphine accumulate in the interbacular spaces (enclosures) of the exine (Figs. 2. 16 and 2.17). The lipid fractions remain on the surface of the tectum sealing the foot layer as the grain contracts on desiccation. In many respects tryphine is similar to Pollenkitt.

The biological function of this material is uncertain. Some botanists suggest that it may act to attract insects, protect against ultraviolet rays or ensure attachment to insect bodies.

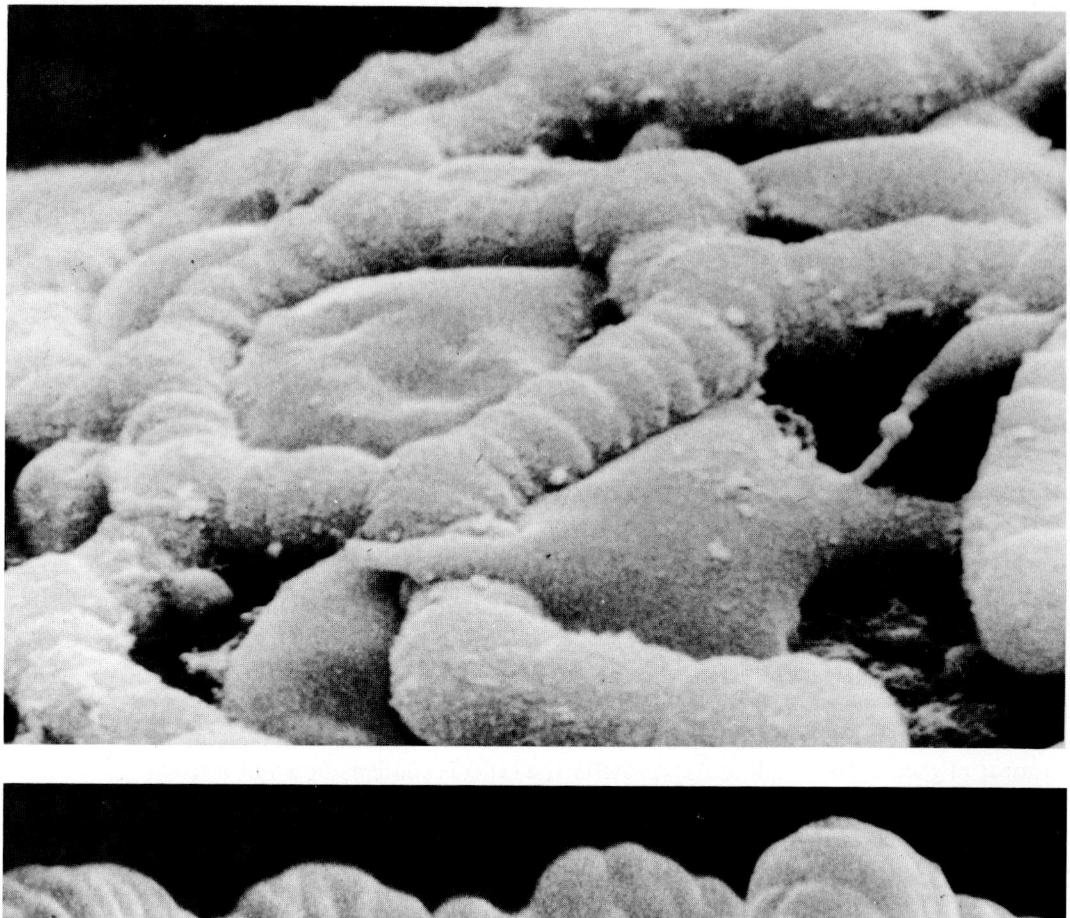

Fig. 2.16 Mature pollen grain of Easter lily (*Lilium longiflorum*).

Top: Outer surface is sealed by lipoidal material (tryphine). The area enclosed by the bacula corresponds to the outer surface of the nexine. ×9300. Bottom: Sporopollen in globules are present on the surface of the enclosures. All the heads are in contact with each other. ×9300.

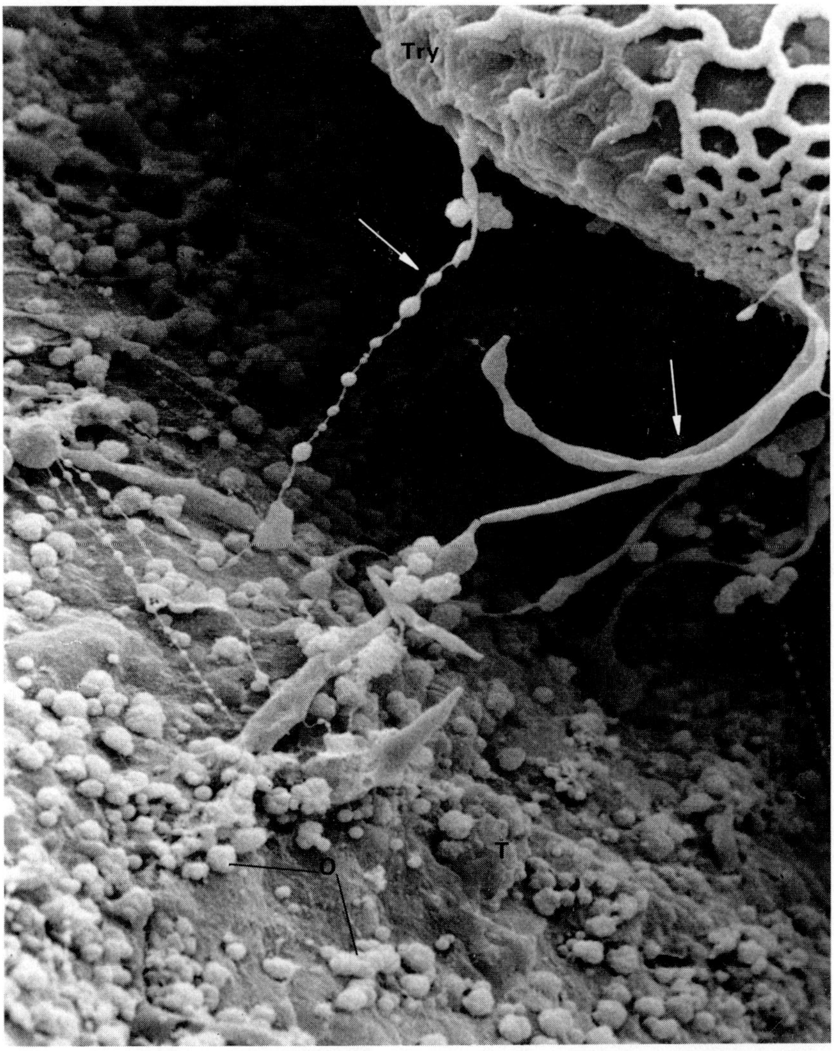

Fig. 2.17 Cross section of anther locule of Easter lily two days before anthesis, showing structural relationships between pollen grain and tapetum. The orbicules and ropelike structures (arrows) are present on the locular face of the tapetum. Pollen grain with adhering ropelike structures is seen in the upper right corner. Note that the tryphine formation is clearly visible on the surface of the enclosure. ×1500. T: tapetum, O: orbicule, Try: tryphine.

2.6.3 Perine

The terms perine and perispore have been imprecisely used in the palynological literature in general. Erdtman defined the exine as "the outer, very resistant layer of the pollen (or spore) wall" and the perine or perispore as "the outermost layer, outside of the exine, in certain spores." The physicochemical qualities of the perine differ more or less from those of the exine.

In preparing pollen samples, our experience, albeit limited, provides definite evidence that an outer layer of the exine occasionally cracks and peels in thin flakes. We believe this observation fits the general perine morphology pattern of pollen grain. Furthermore, the perine has been removed by sufficient treatment with sodium hydroxide to expose the exine for scanning electron or light microscopy.

In *Mirabilis* the perine is observed after enlargement of the exine. An example is the pollen grain of the four-o'clock (*Mirabilis Jalapa* L.), the surface of which appears in Fig. 2.18. Areas where the perine has been ruptured reveal that it lies over a sculptured exine. The perine adheres tightly to the exine and possesses ornamentation of spinules that are exactly like those which make up the exine.

Some pollen grains show an irregularly bumpy sculptine appearance (e.g. *Malvaceae*). When pollen grains from the brazilian abutilon (*Abutilon megapotamicum*) are assayed with sodium hydroxide, the deposition over the sculptured exine is perinous (Fig. 2.19). In usual pollen one can

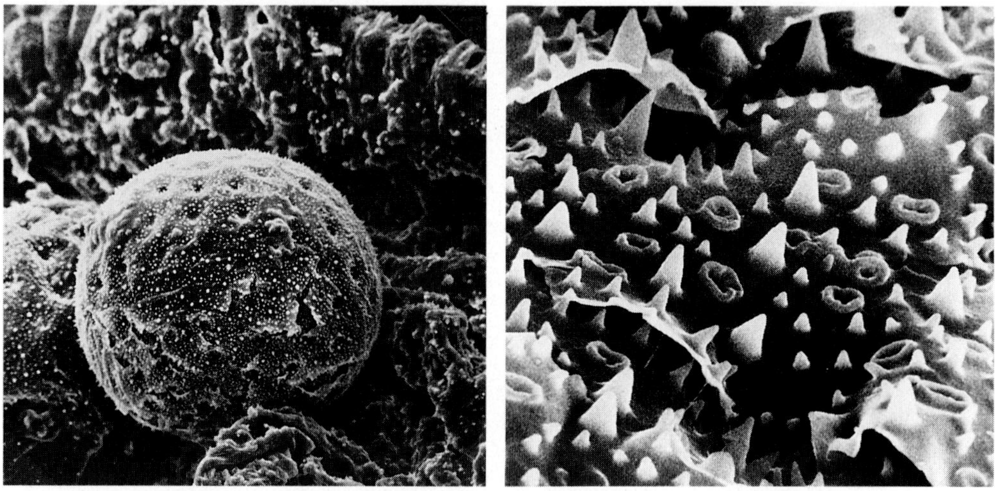

Fig. 2.18 Four-o'clock pollen grain is magnified 360 times (*left*); it displays the external appearance of a perinous pollen grain within the anther locule. Detail of perine surface, magnified 3600 times (*right*). Note that an outer layer of the exine peels in thin flakes.

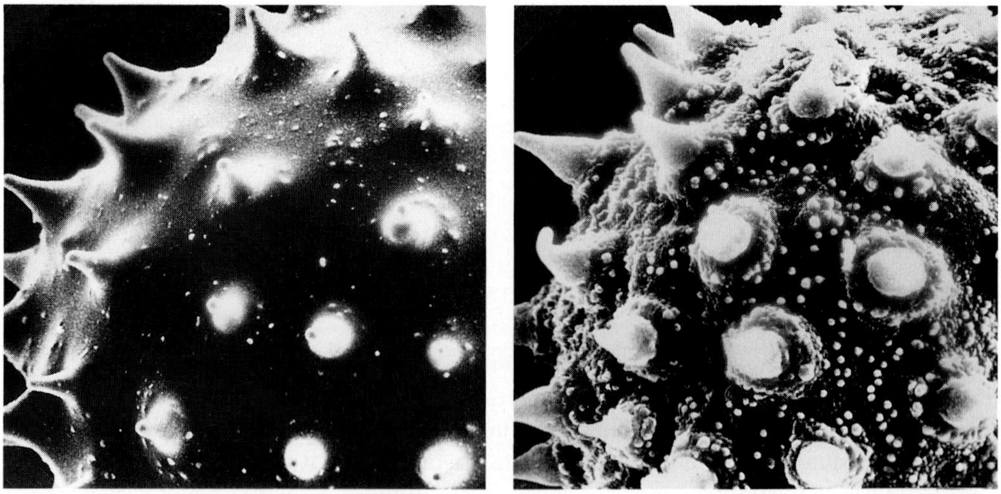

Fig. 2.19 Mature pollen grain of brazilian abutilon is magnified 2160 times (*left*); it is covered with long spines. Note relatively smooth surface of exine. Granular surface of the exine, after removal of the perine layer by hot alkali treatment (*right*). ×2160.

see long spines and a slightly granular exine surface, while in alkali-treated pollen the spines generally have a raised, prominent base and the surface is apparently more granular. Thus, ornamentation of the pollen of this species is due solely to deposition of the perine.

The ontogeny of the perine in these pollen has not yet been studied, although true perine is generally assumed to be derived from the tapetum.

2.6.4 Viscin threads

In *Onagraceae*, numerous thread-like structures called viscin threads develop in pollen grains (Fig. 2.20). Viscin threads hold together large numbers of pollen grains from dehisced anthers (Fig. 2.20 A). Because viscin threads hold numbers of pollen grains together it has been assumed that they play an important role in pollination. They are also observed in *Rhododendron.*

For example, the sample of evening primrose (*Oenothera parviffora*) had globular sculptural

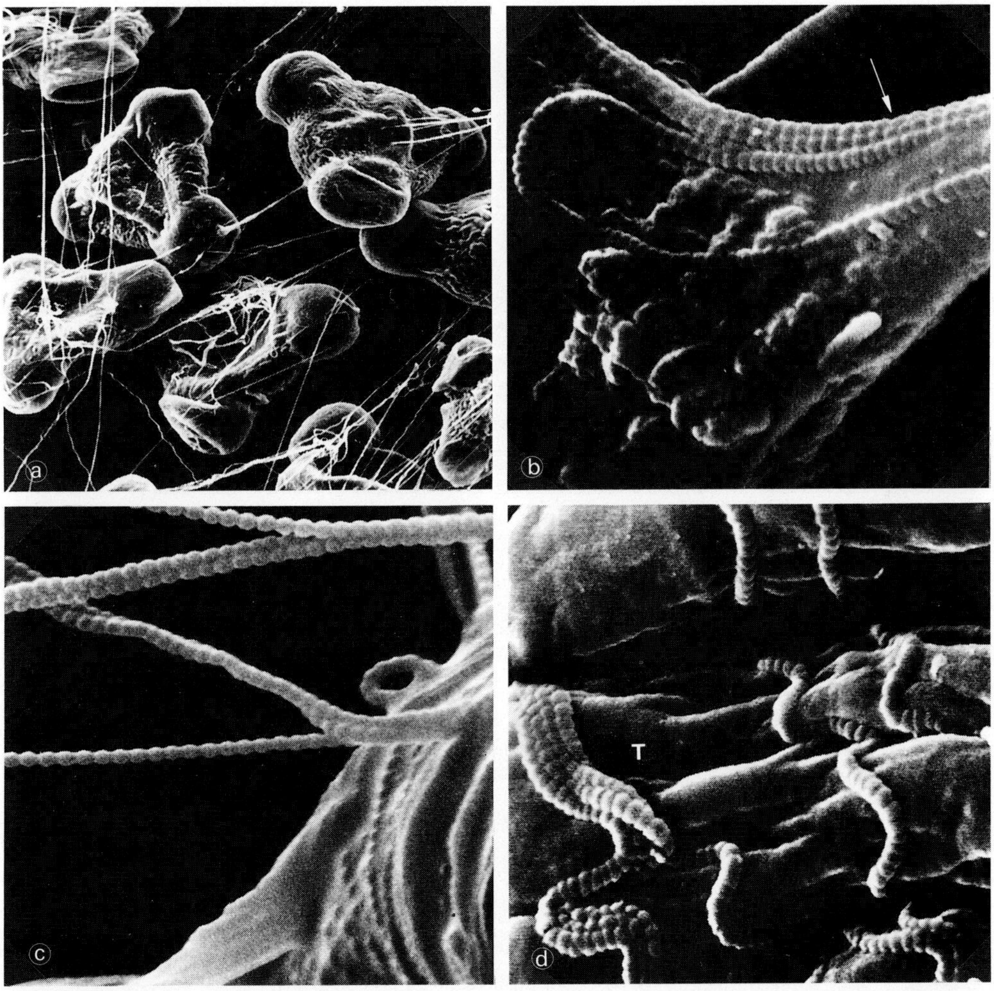

Fig. 2.20 SEM micrographs of evening primrose pollen.
A: Group of pollen with viscin threads. ×720. B: SEM micrograph showing both lightly segmented viscin threads (arrow) and a portion of thread attached to the exine. ×7200. C: SEM micrograph showing threads on proximal polar face. Note attachment of thread bases to exine surface. ×7200. D: Groups of segmented threads over a surface of tapetum. ×7200. T: tapetum.

elements, although on the proximal surface there are also elongated elements associated with the viscin threads (Fig. 2.20 C). Viscin threads are essentially simple, although they often form twisted strands (Fig. 2.20 C). Threads often arise in clumps and fuse along their lengths (Fig. 2.20 B). They vary greatly in thickness and exhibit a beaded or stacked disc organisation of sporopollenin. Also, the viscin threads can be seen to be best developed in the inner locular surface (or tapetum) at anthesis (Fig. 2.20 D).

Morphological variation of the threads is notable. Two main types of viscin thread morphology are recognized, a smooth pattern (Fig. 2.21) and a beaded design (Fig. 2.20). We also observed a smooth type of viscin thread in rose-bay (*Rhododendron Metternichii*), which has pollen in permanent tetrahedral tetrads at maturity (Fig. 2.21 A). Tectum surface showed randomly oriented rods of sporopollenin (Fig. 2.21 C). Pollen grains are joined by numerous long viscin threads to form large clumps. Viscin threads are multiple and branching and fuse several times near the grain (Fig. 2.21 B). The viscin threads are attached directly to the exine surface (Fig. 2.21D).

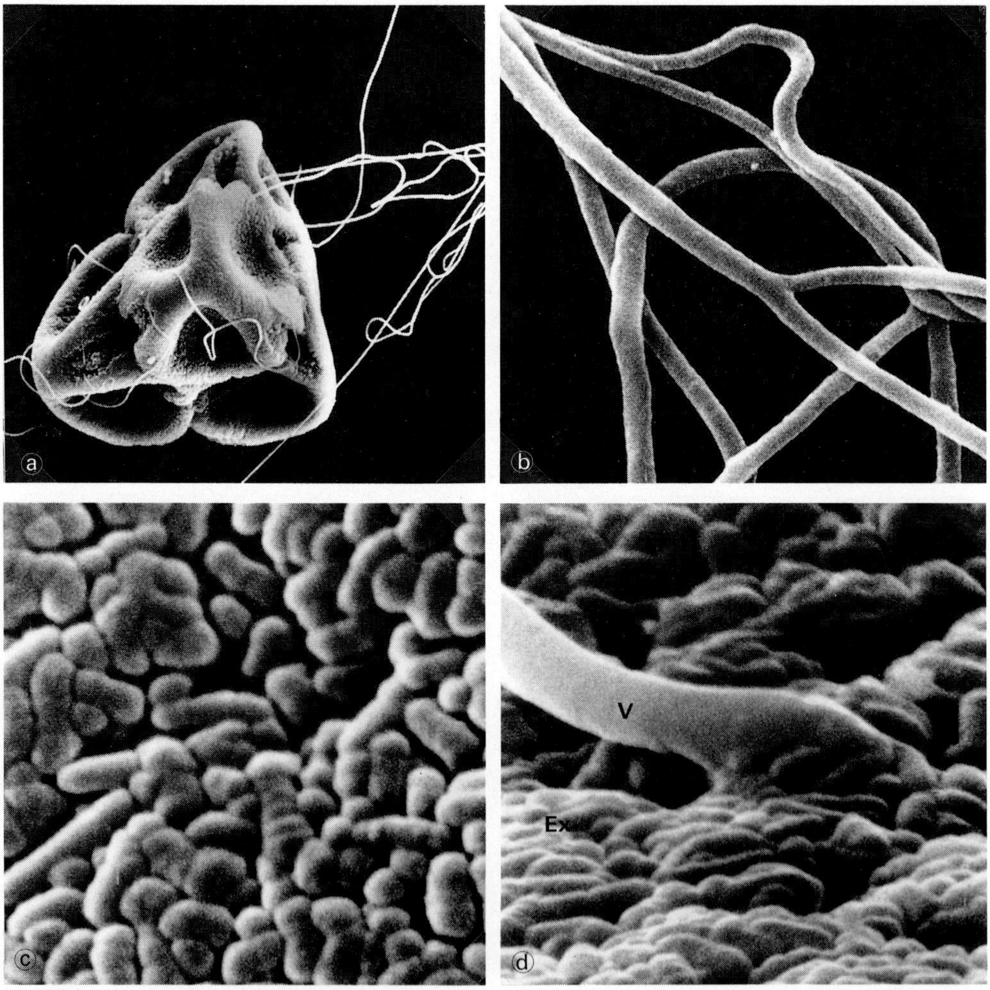

Fig. 2.21 SEM micrographs of rose-bay pollen.
A: Ropelike structures are viscin threads. ×720. B: SEM micrograph showing smooth, branched viscin threads. ×7200. C: Exine surface of the pollen grain showing randomly oriented rods of sporopollenin. ×21600. D: SEM micrograph showing a portion of attachment of viscin thread base to exine surface. ×21600. Ex: exine, V: viscin thread.

Implicit in the definition of viscin threads is that one end is attached to the exine while the other end is free.

2.6.5 Exinal connections

Thread-like structures are also observed on pollen grains of some *Leguminosae*. These are usually limited to one or two per pollen grain, but some grains do not have the structures. The ends of the thread-like structures are attached to the reticulate surface of different pollen grains. In view of the absence of "free ends," the structures are more appropriately considered to be exinal connections. The exinal connections of spindle-tree (*Euonymus alata*) are similar to those of the legumes (Fig. 2.22). Each end is attached to the highly reticulate exine surface of different grains. In the spindle tree the area of the stigma is relatively small to the area of pollen. Therefore, pollination is facilitated by short connections which hold small numbers of pollen grains together.

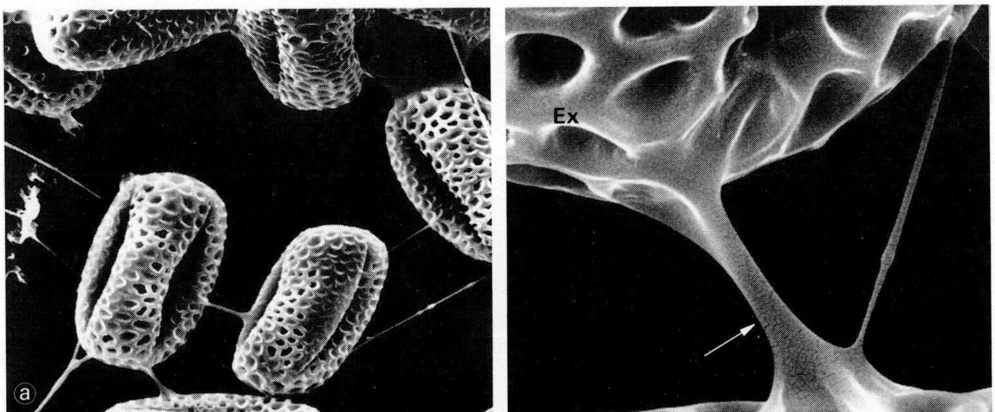

Fig. 2.22 Mature pollen of spindle-tree.
A: Pollen grains joined by exinal connections. ×720. B and C: Pollen grains showing continuity of exine and connections (arrow). B: ×5400, C: ×2160. D: A higher magnification of exinal connection of the same pollen grain as in C showing its beautiful design. ×10800. Ex: exine.

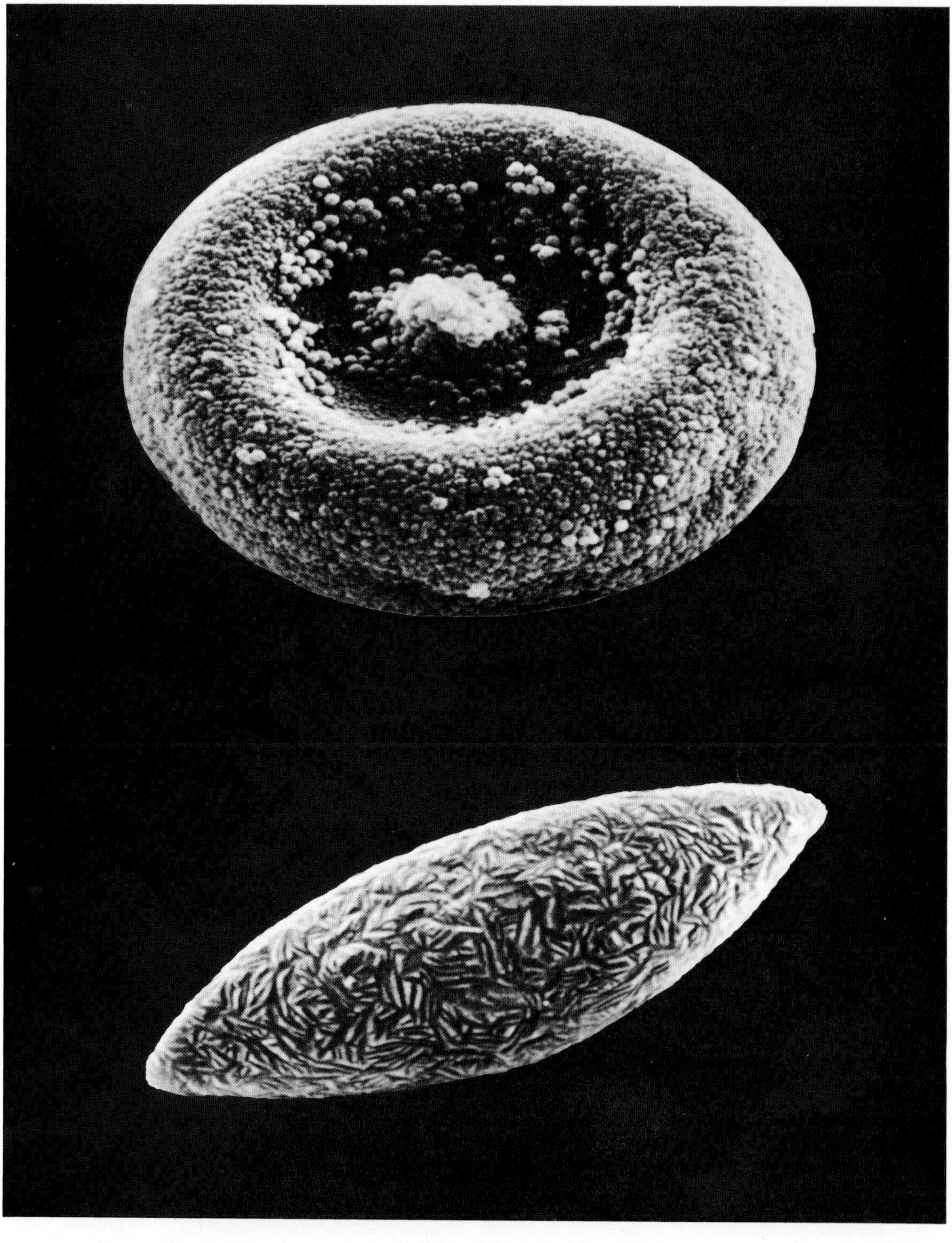

Fig. 2.23 Pollen morphology (1).
Top: Japanese Red Cedar (*Crypomeria japonica* D.Don) [**Taxodiaceae**] (×2700).
Bottom: Maidenhair Tree (*Ginkgo biloba* L.) [**Ginkgoaceae**] (×2700).

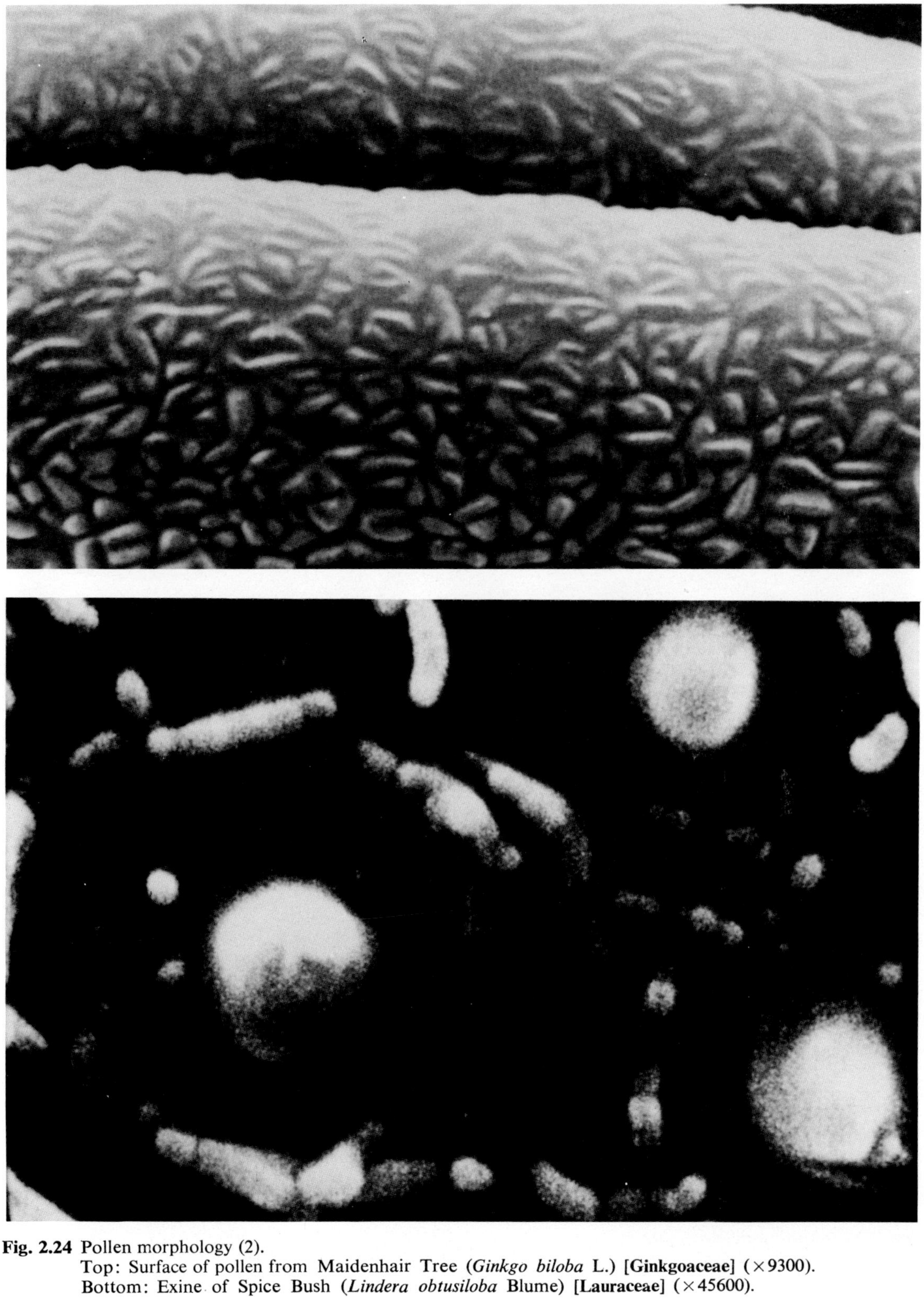

Fig. 2.24 Pollen morphology (2).
Top: Surface of pollen from Maidenhair Tree (*Ginkgo biloba* L.) [**Ginkgoaceae**] (×9300).
Bottom: Exine of Spice Bush (*Lindera obtusiloba* Blume) [**Lauraceae**] (×45600).

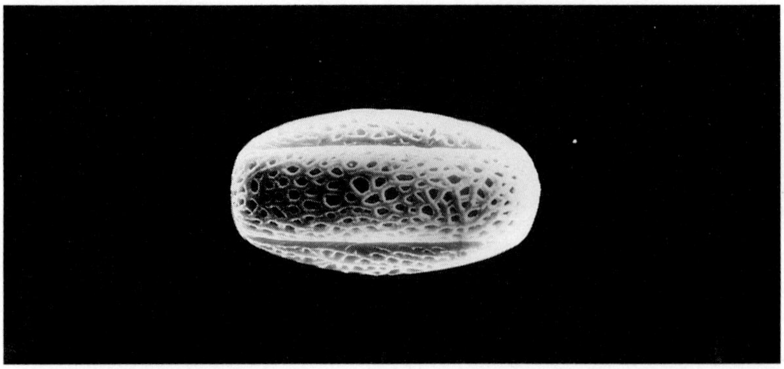

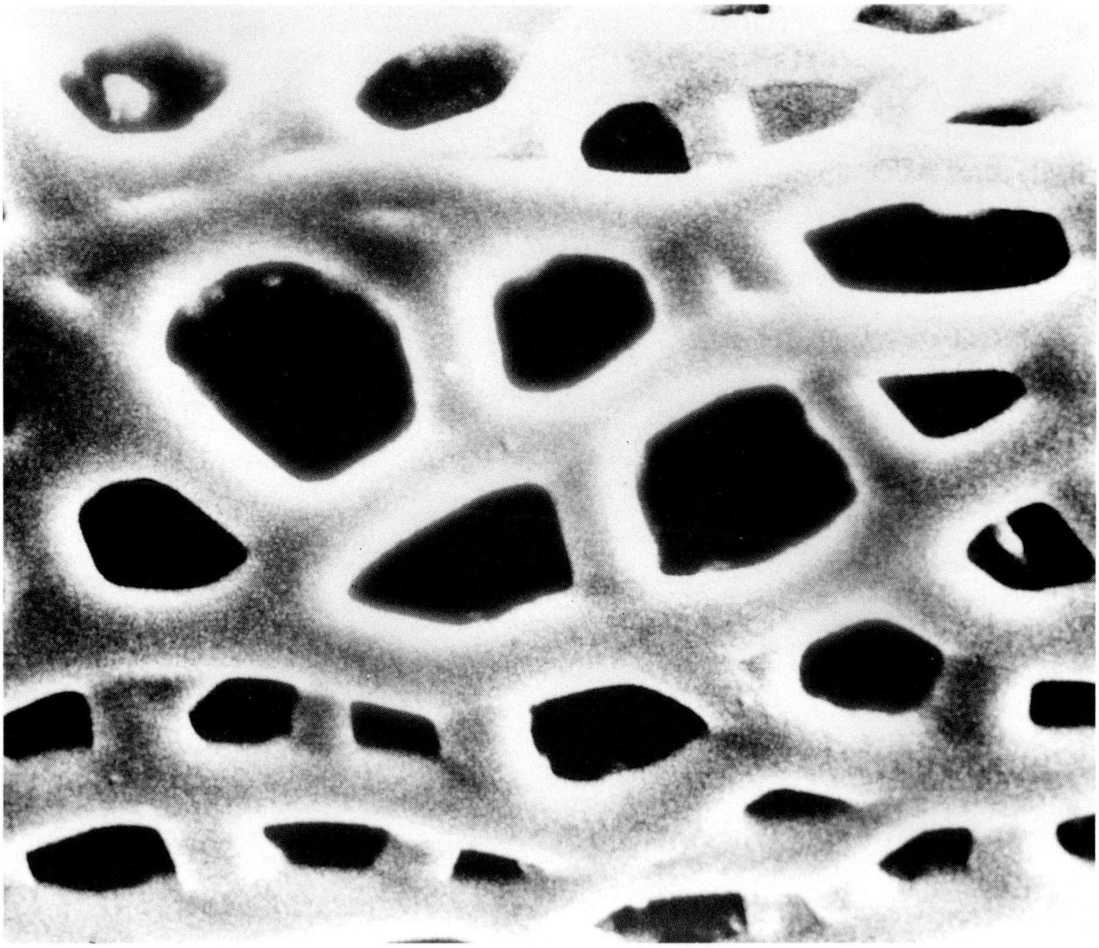

Fig. 2.25 Pollen morphology (3).
Top: *Trochodendron aralioides* Sieb. et Zucc. [**Trochodendraceae**] (×1600).
Bottom: Detail of ornamentation (×25000).

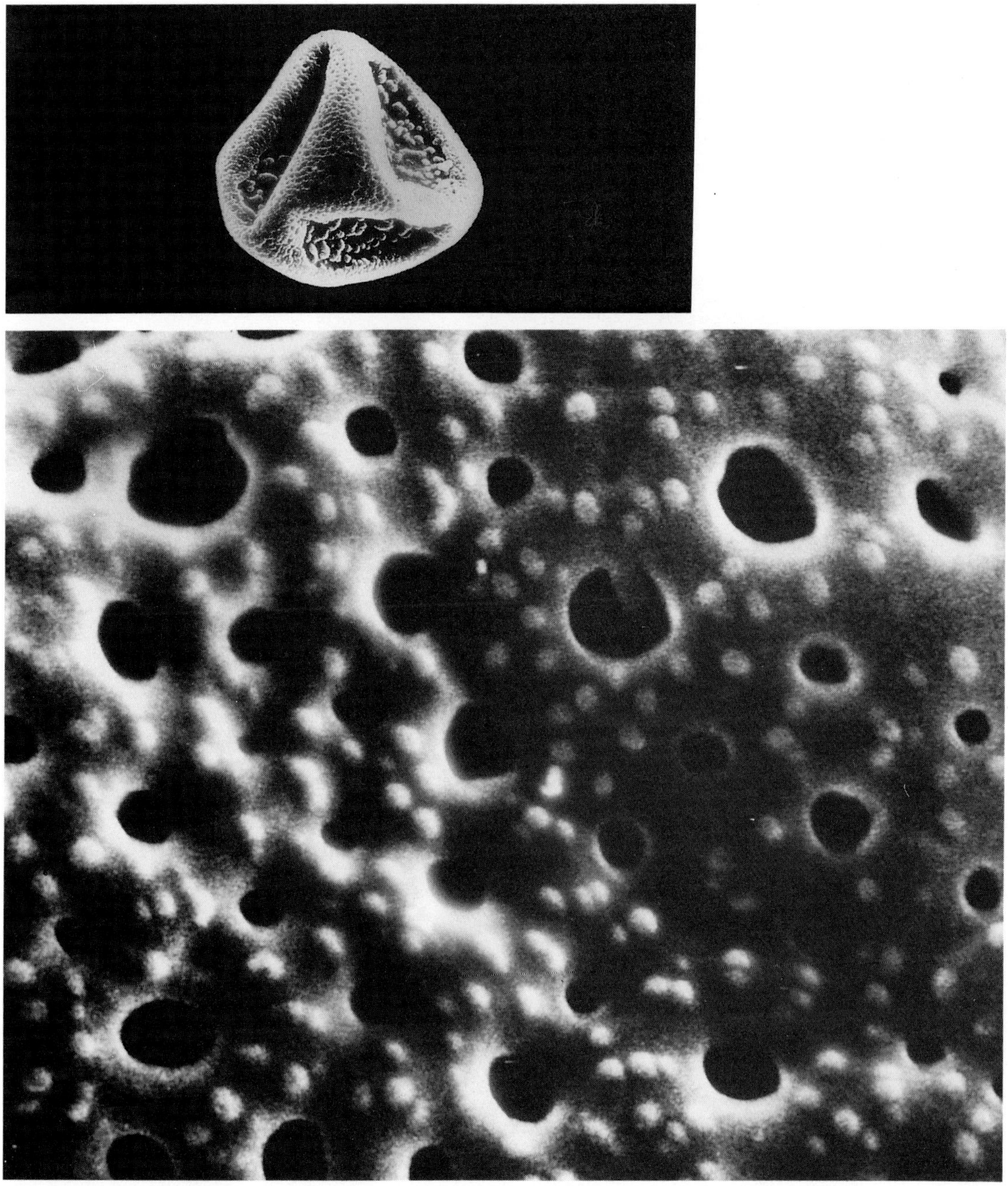

Fig. 2.26 Pollen morphology (4).
Top: *Euptelea polyandra* Sieb. et Zucc. [**Eupteleaceae**] ($\times$1300).
Bottom: Detail of ornamentation ($\times$19500).

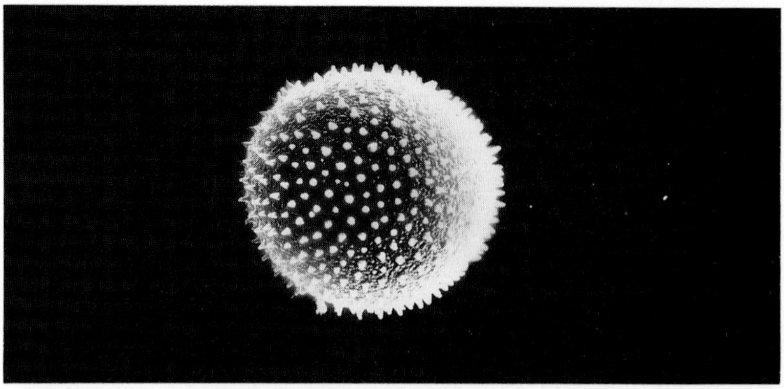

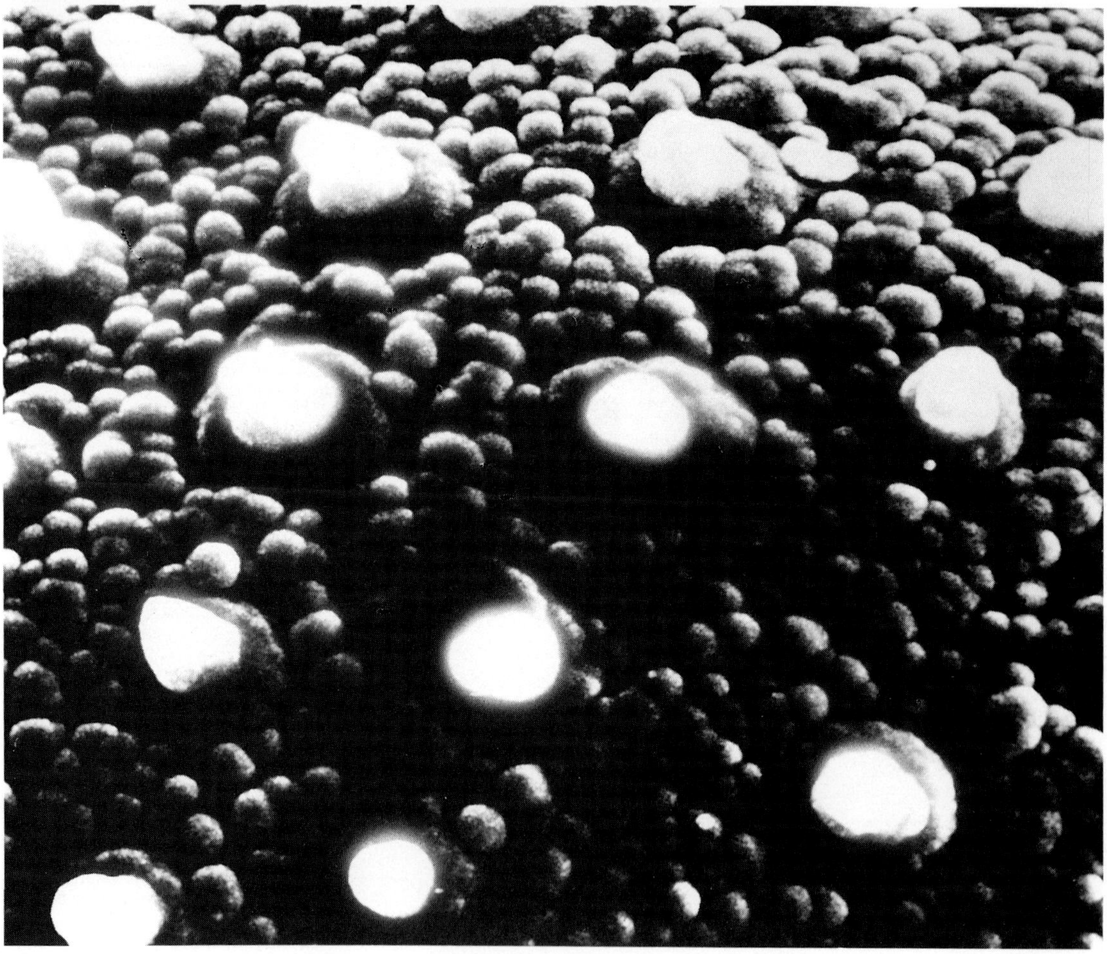

Fig. 2.27 Pollen morphology (5).
Top: Camphor tree (*Cinnamomum camphore* Sieb.) **[Lauraceae]** (×1625).
Bottom: Detail of ornamentation (×24400).

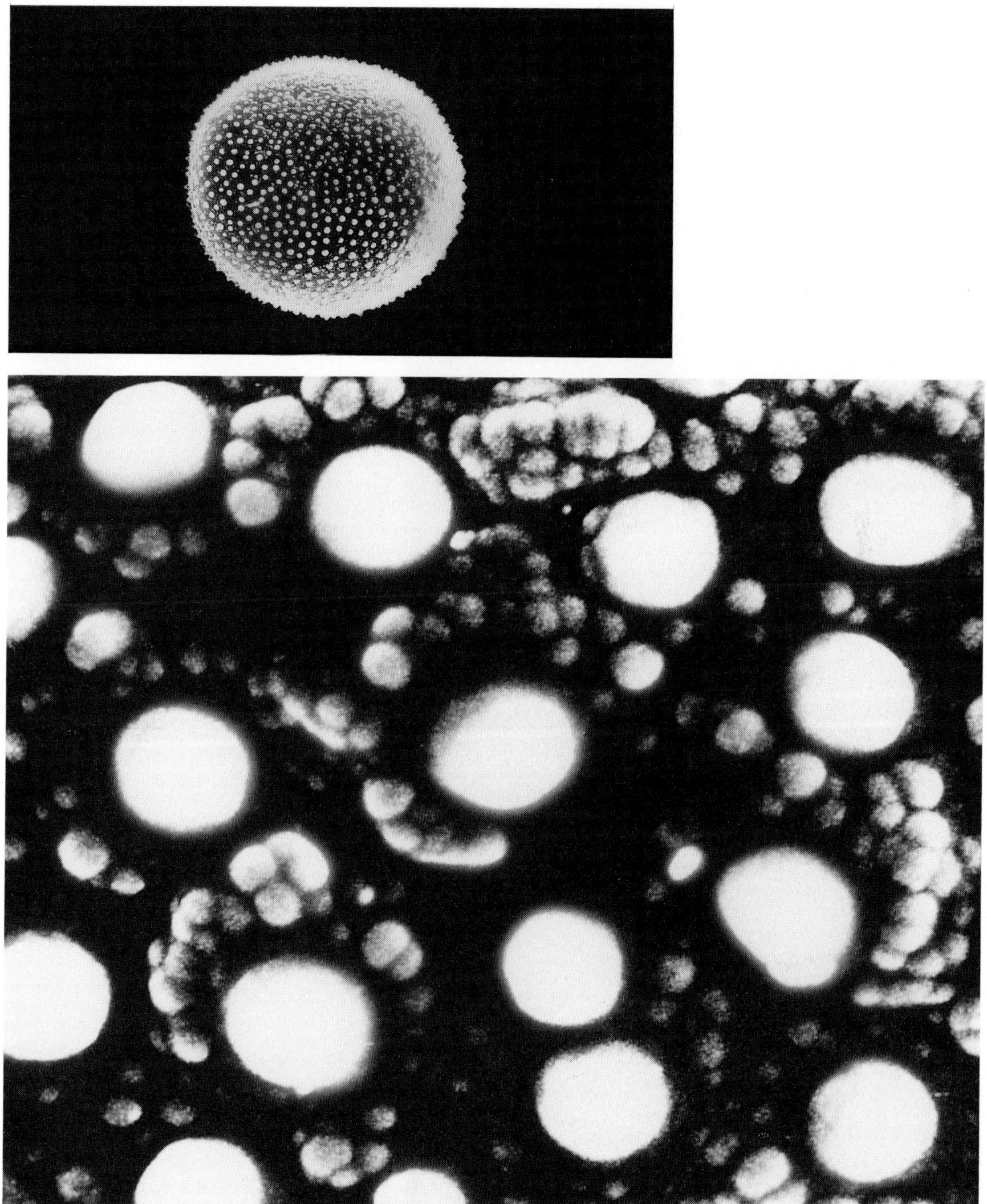

Fig. 2.28 Pollen morphology (6).
Top: *Machilus thunbergii* Sieb. et Zucc. [**Lauraceae**] (×1300).
Bottom: Detail of ornamentation (×24400).

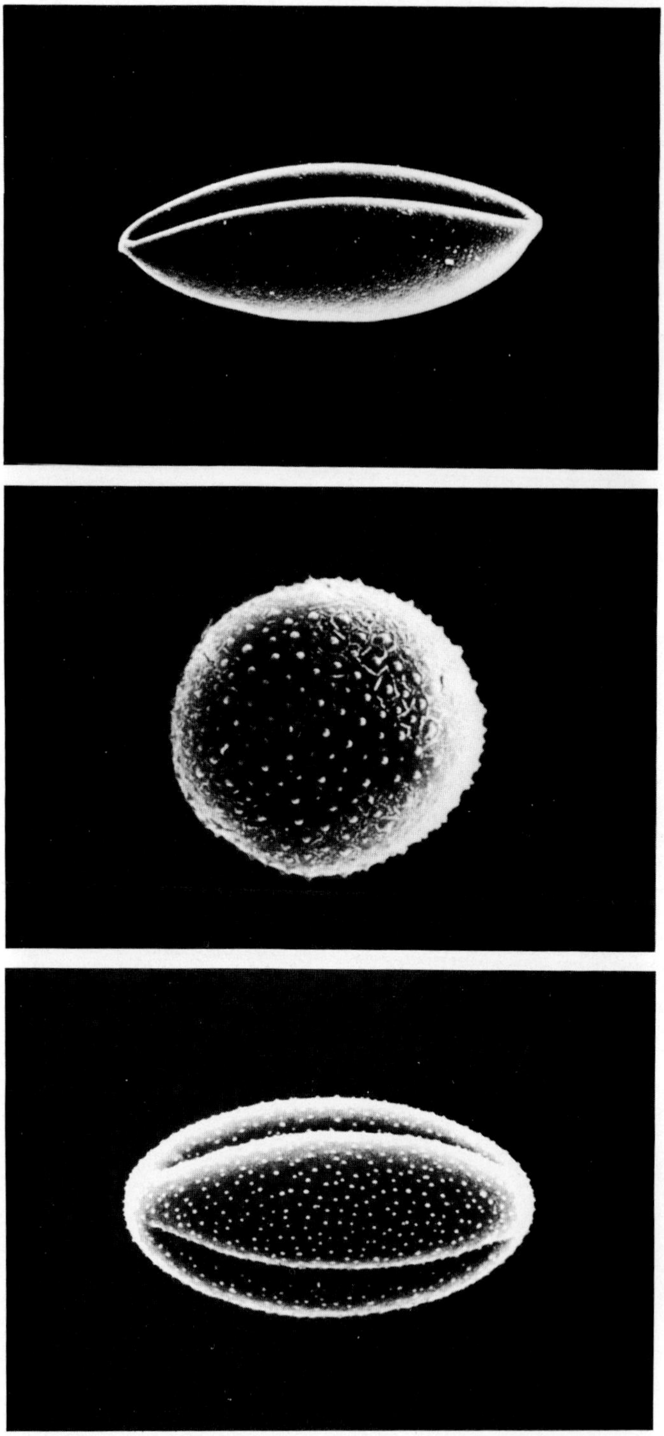

Fig. 2.29 Pollen morphology (7).
Top: Japanese Big-leaf Magnolia (*Magnolia obovata* Thunb.) [**Magnoliaceae**] (×600).
Middle: Spice Bush (*Lindera obtusiloba* Blume) [**Lauraceae**] (×1700).
Bottom: *Pulsatilla cernua* Spreng. [**Ranunculacea**] (×1200).

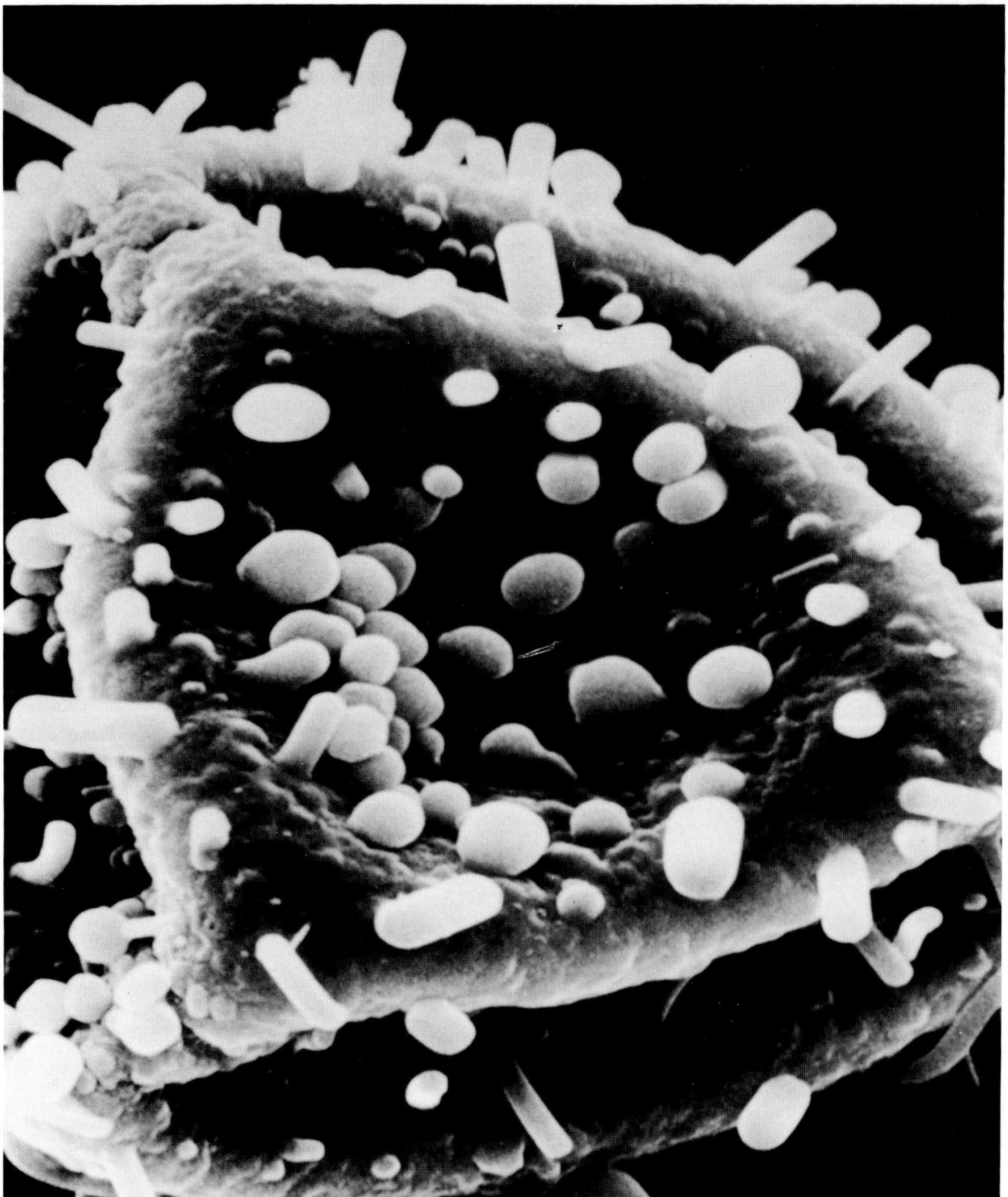

Fig. 2.30 Pollen morphology (8).
Surface of pollen from Waterlily (*Nymphaea odorata* Ait. var. *rosea* Pursh) [**Nymphaeaceae**] (×5600).

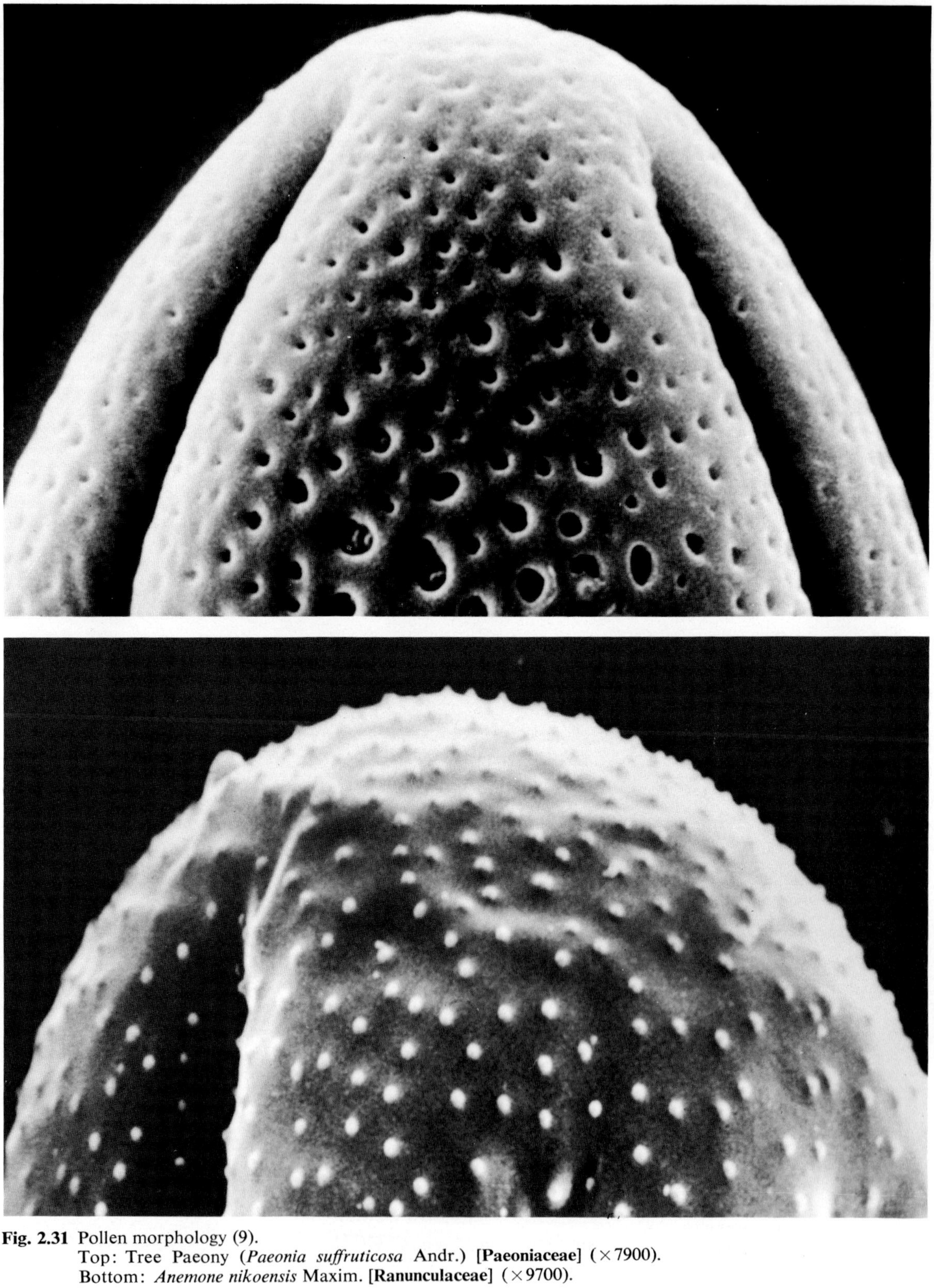

Fig. 2.31 Pollen morphology (9).
Top: Tree Paeony (*Paeonia suffruticosa* Andr.) [**Paeoniaceae**] (×7900).
Bottom: *Anemone nikoensis* Maxim. [**Ranunculaceae**] (×9700).

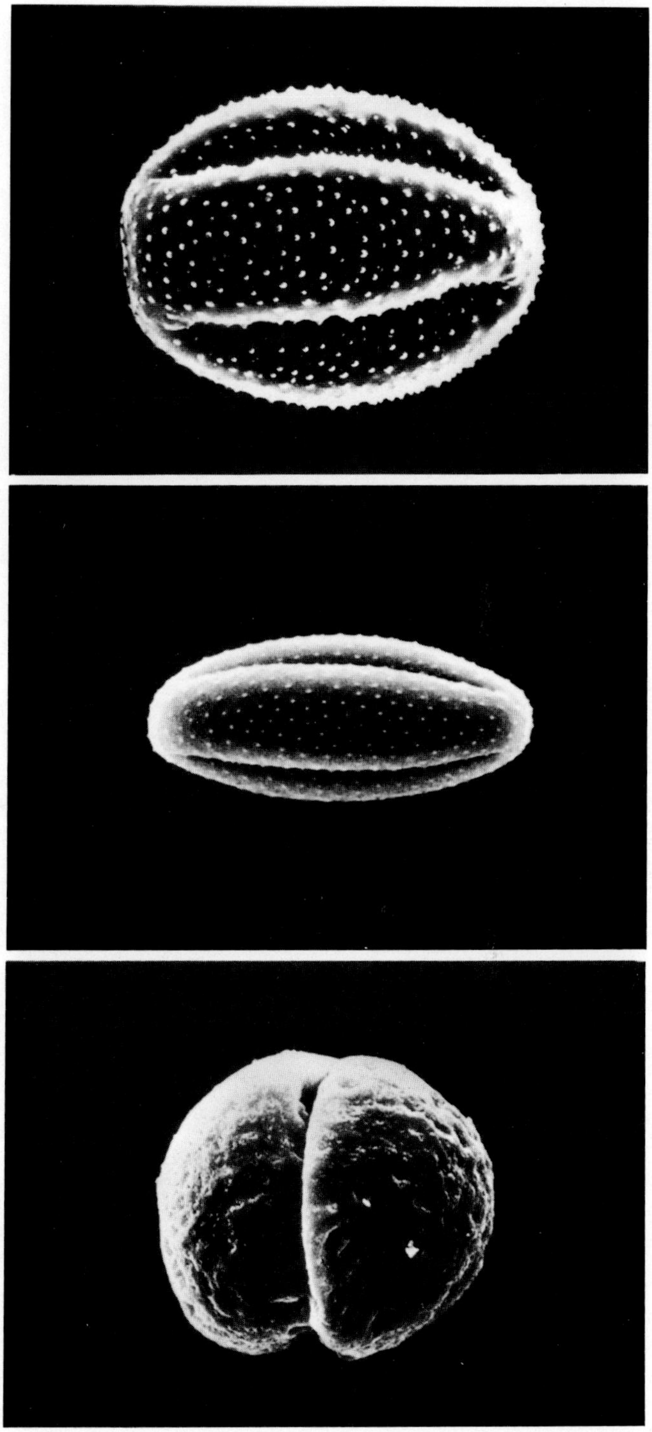

Fig. 2.32 Pollen morphology (10).
Top: *Anemone stolonifera* Maxim. **[Ranunculaceae]** (×1700).
Middle: Columbine (*Aquilegia buergeriana* Sieb. et Zucc.) **[Ranunculaceae]** (×1700).
Bottom: *Houttuynia cordata* Thunb. **[Saururaceae]** (×2500).

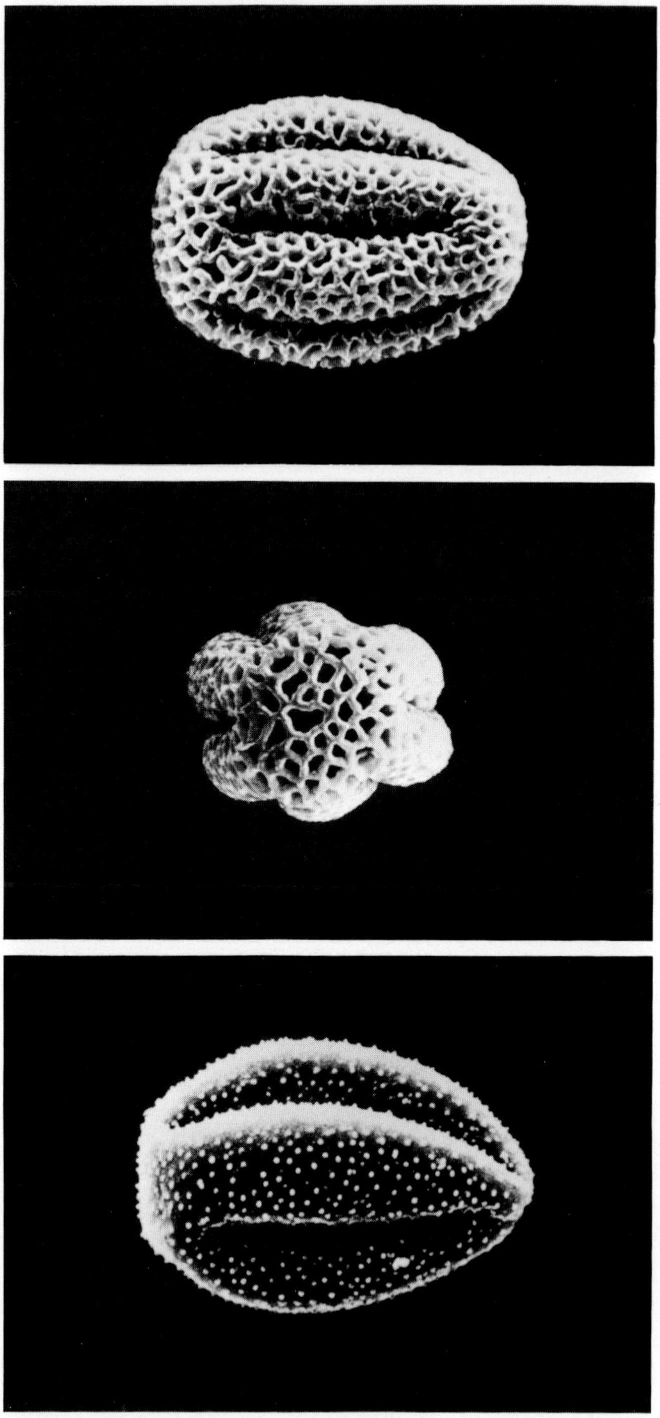

Fig. 2.33 Pollen morphology (11).
Top: Equatorial view of *Chloranthus japonicus* Sieb. [**Chloranthaceae**] (×1700).
Middle: Polar view of *Chloranthus japonicus* Sieb. (×1900).
Bottom: Corn-poppy (*Papaver Rhoeas* L.) [**Papaveraceae**] (×1700).

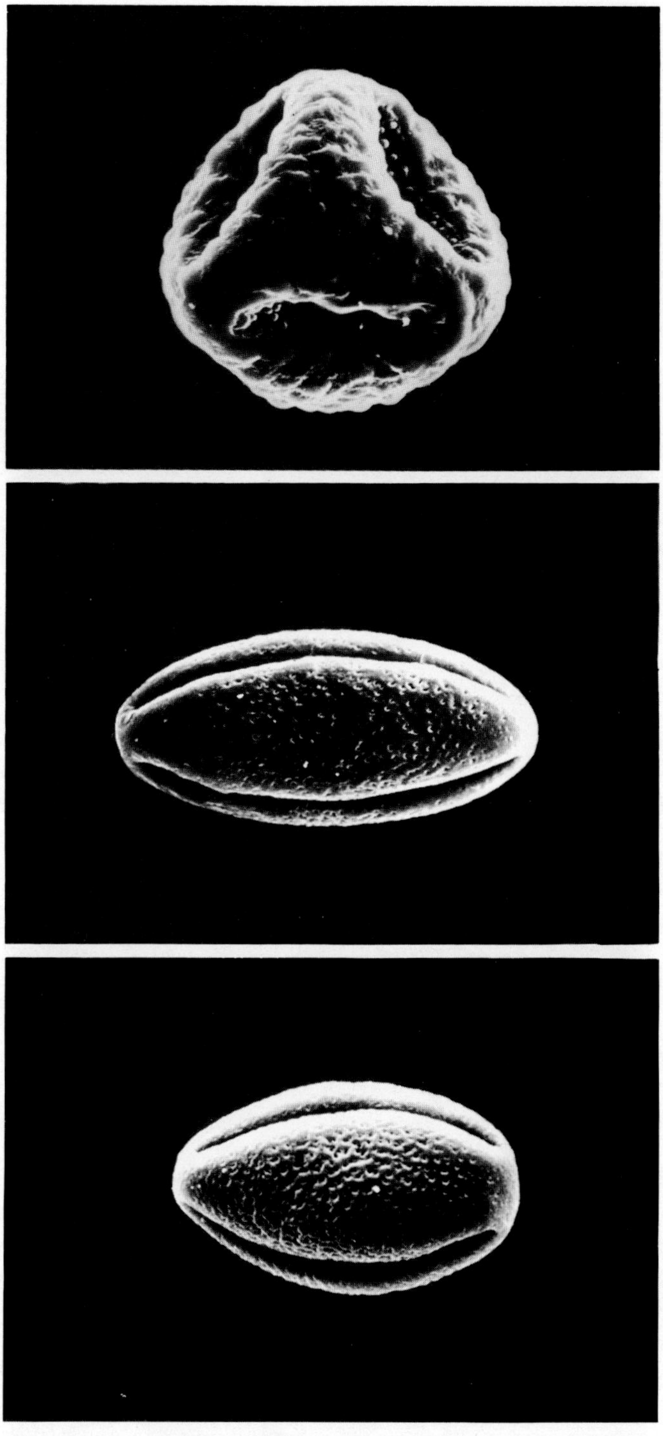

Fig. 2.34 Pollen morphology (12).
Top: Corydalis (*Corydalis ochotensis* Turcz.) [**Fumariaceae**] (×1200).
Middle: Camellia (*Camellia japonica* L.) [**Theaceae**] (×800).
Bottom: Sasanqua Tea (*Camellia sasanqua* Thunb.) [**Theaceae**] (×800).

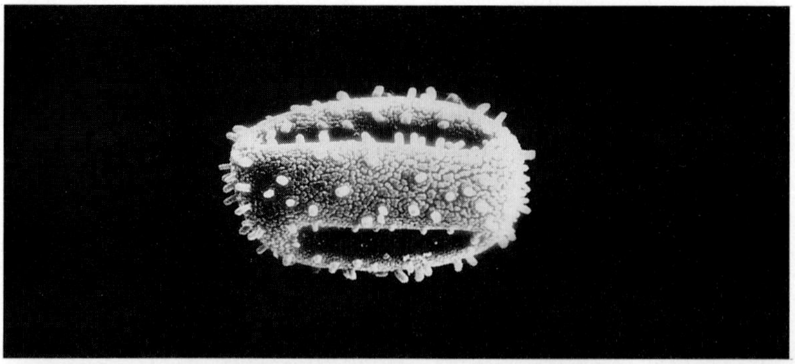

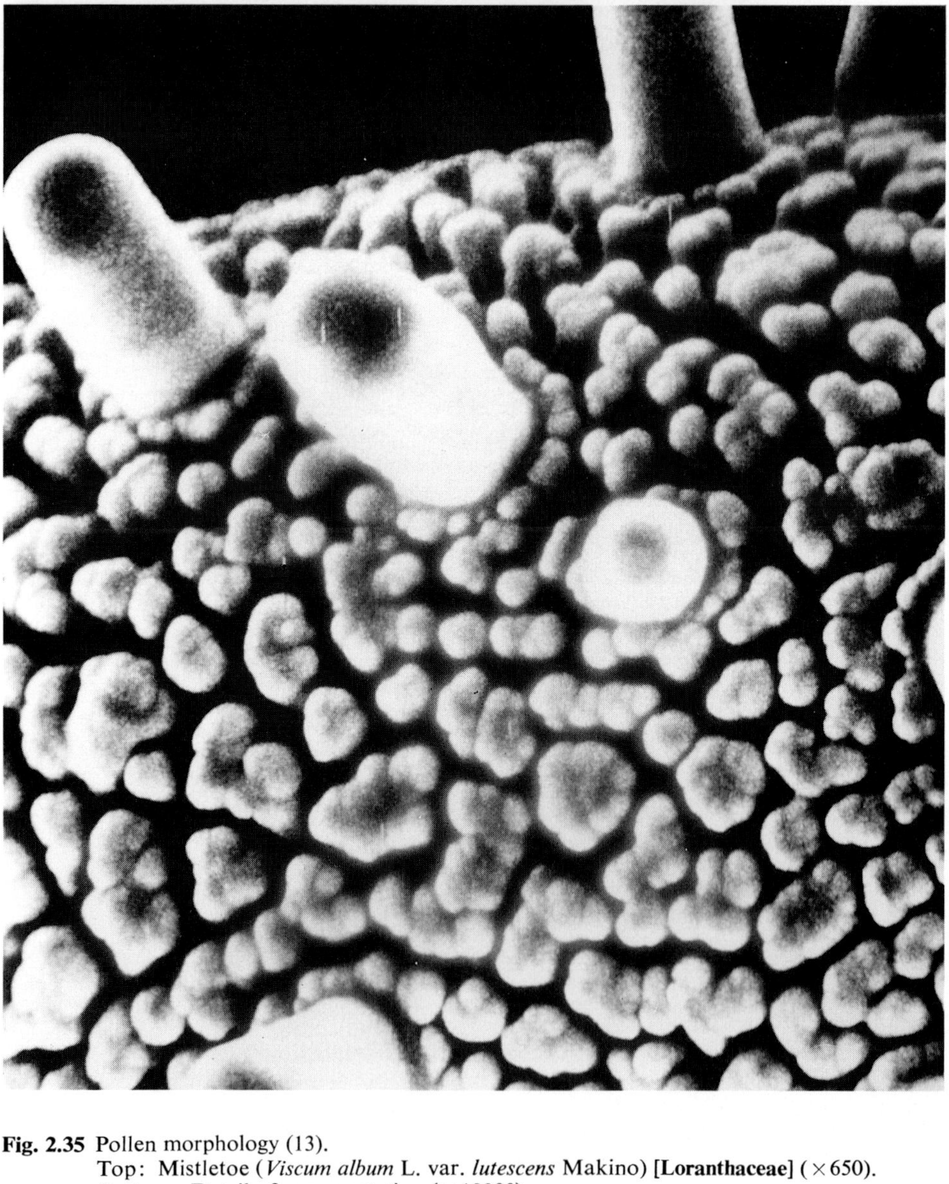

Fig. 2.35 Pollen morphology (13).
Top: Mistletoe (*Viscum album* L. var. *lutescens* Makino) **[Loranthaceae]** (×650).
Bottom: Detail of ornamentation (×12200).

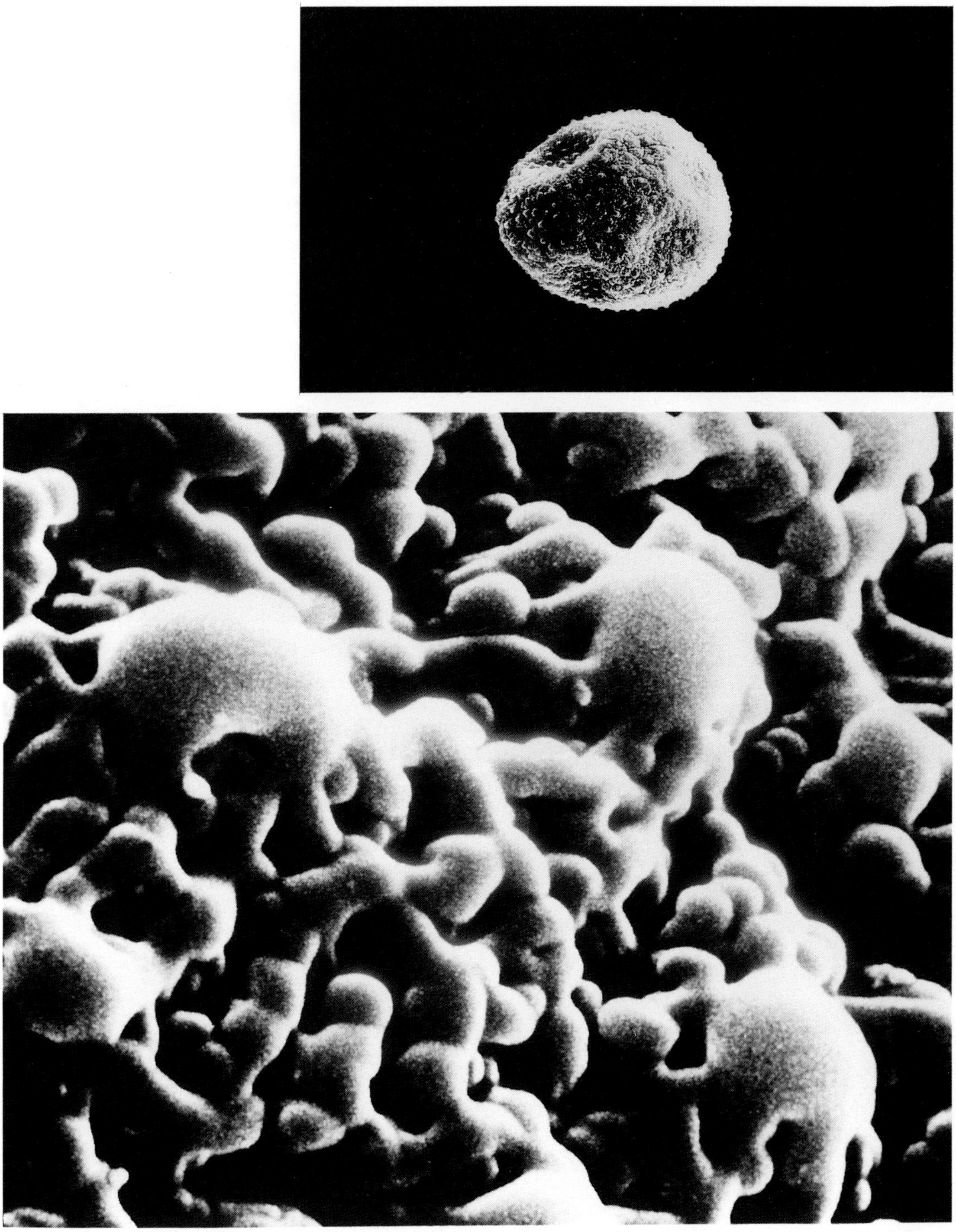

Fig. 2.36 Pollen morphology (14).
Top: Asarabacca (*Asarum caulescens* Maxim.)[**Aristolochiaceae**] (×650).
Bottom: Detail of ornamentation (×24400)

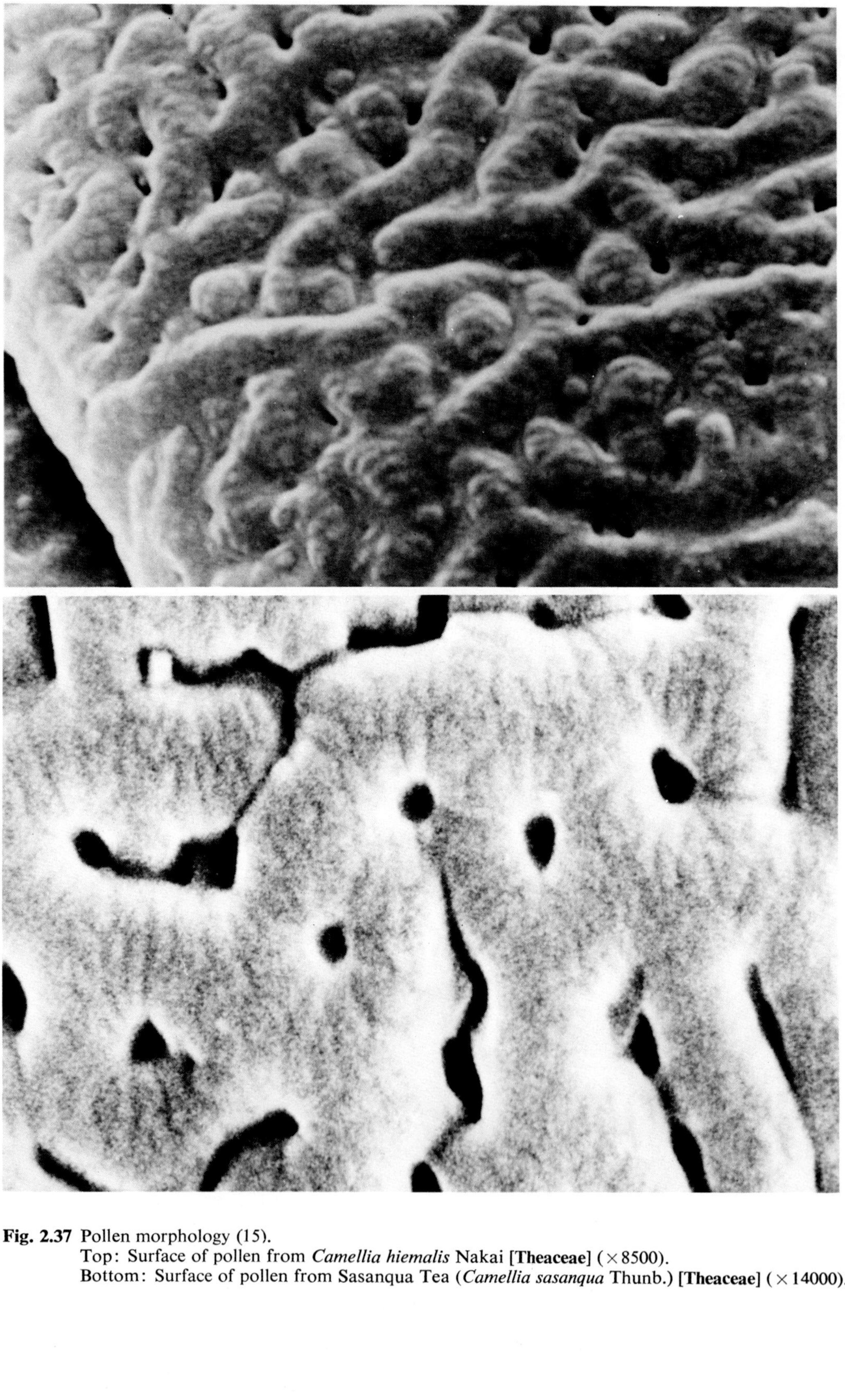

Fig. 2.37 Pollen morphology (15).
Top: Surface of pollen from *Camellia hiemalis* Nakai [**Theaceae**] (×8500).
Bottom: Surface of pollen from Sasanqua Tea (*Camellia sasanqua* Thunb.) [**Theaceae**] (×14000).

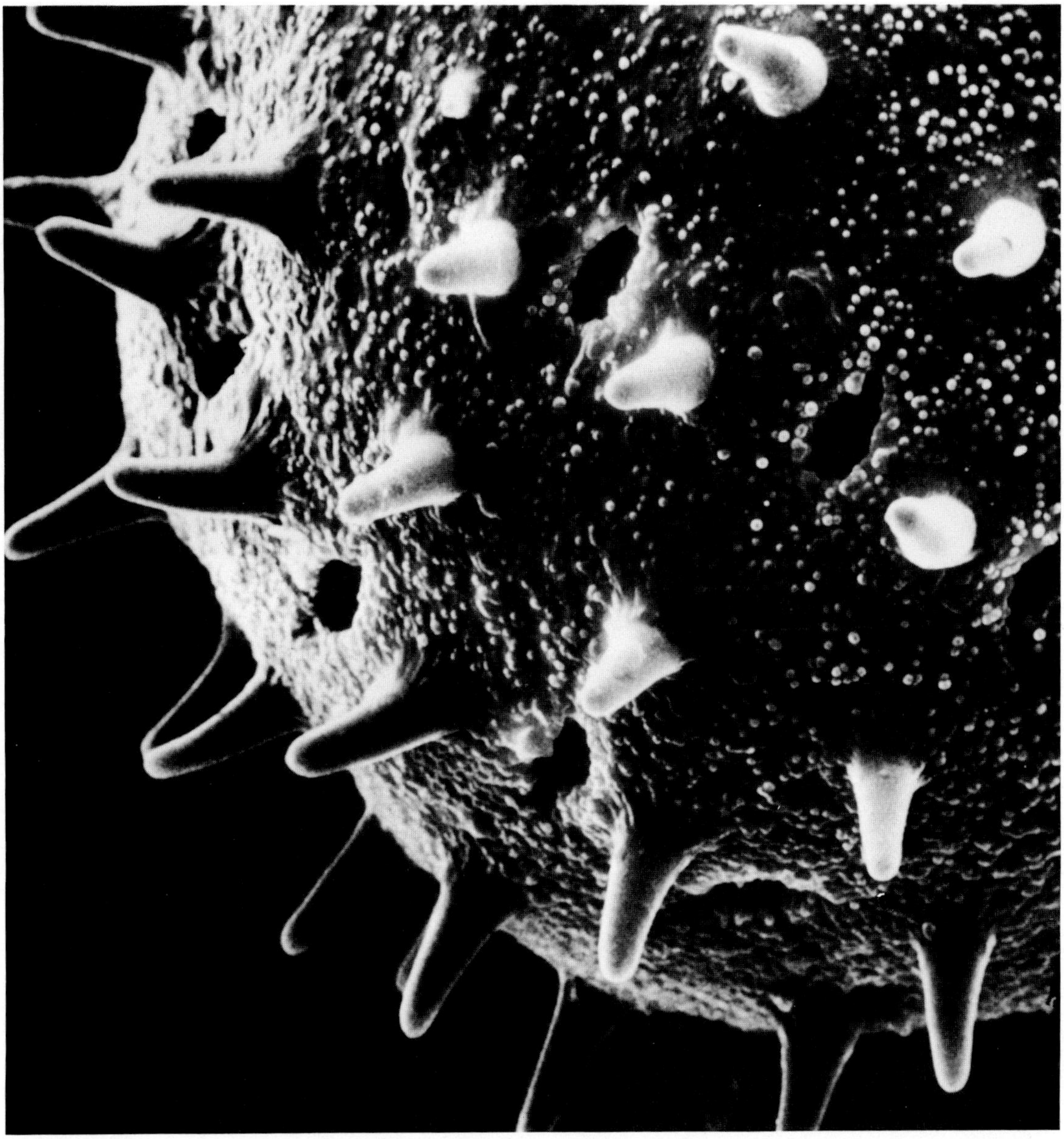

Fig. 2.38 Pollen morphology (16).
Surface of pollen from Rose Mallow (*Hibiscus mutabilis* L.) **[Malvaceae]** (×1700).

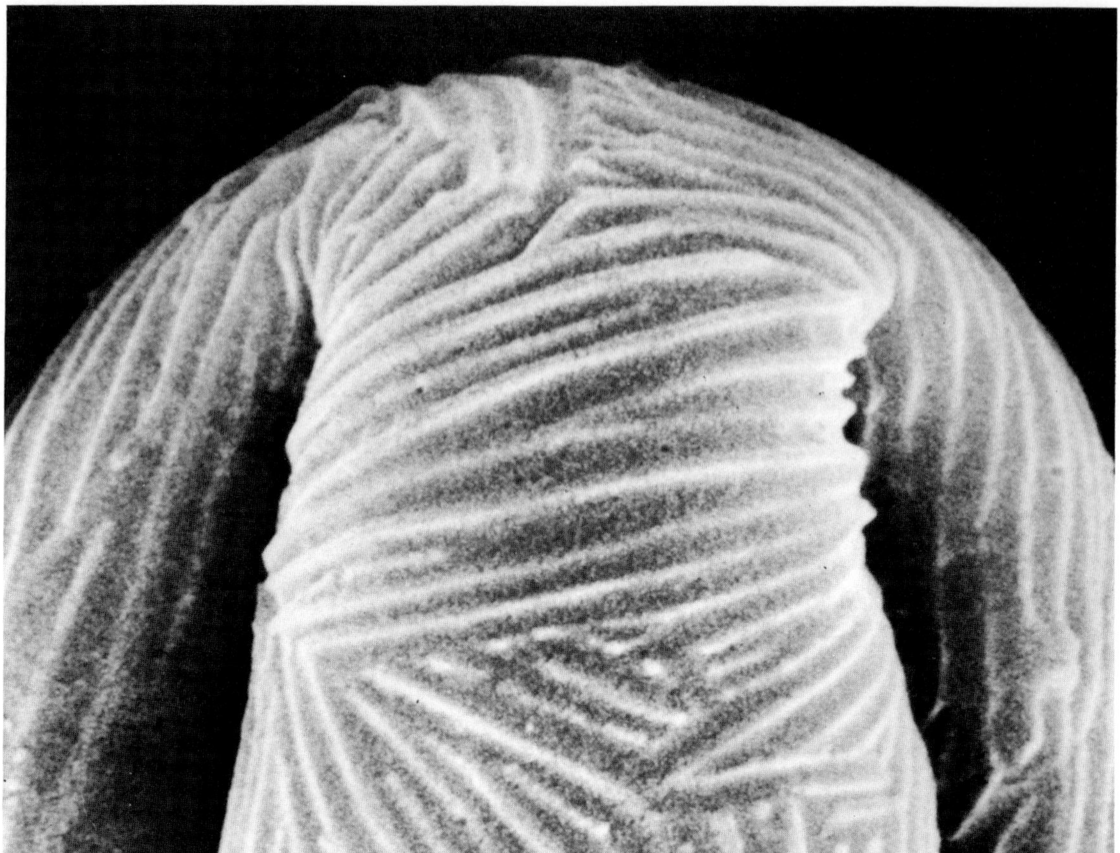

Fig. 2.39 Pollen morphology (17).
Surface of pollen from Begonia (*Begonia Evansiana* Andr.)[**Begoniaceae**]($\times 16000$).

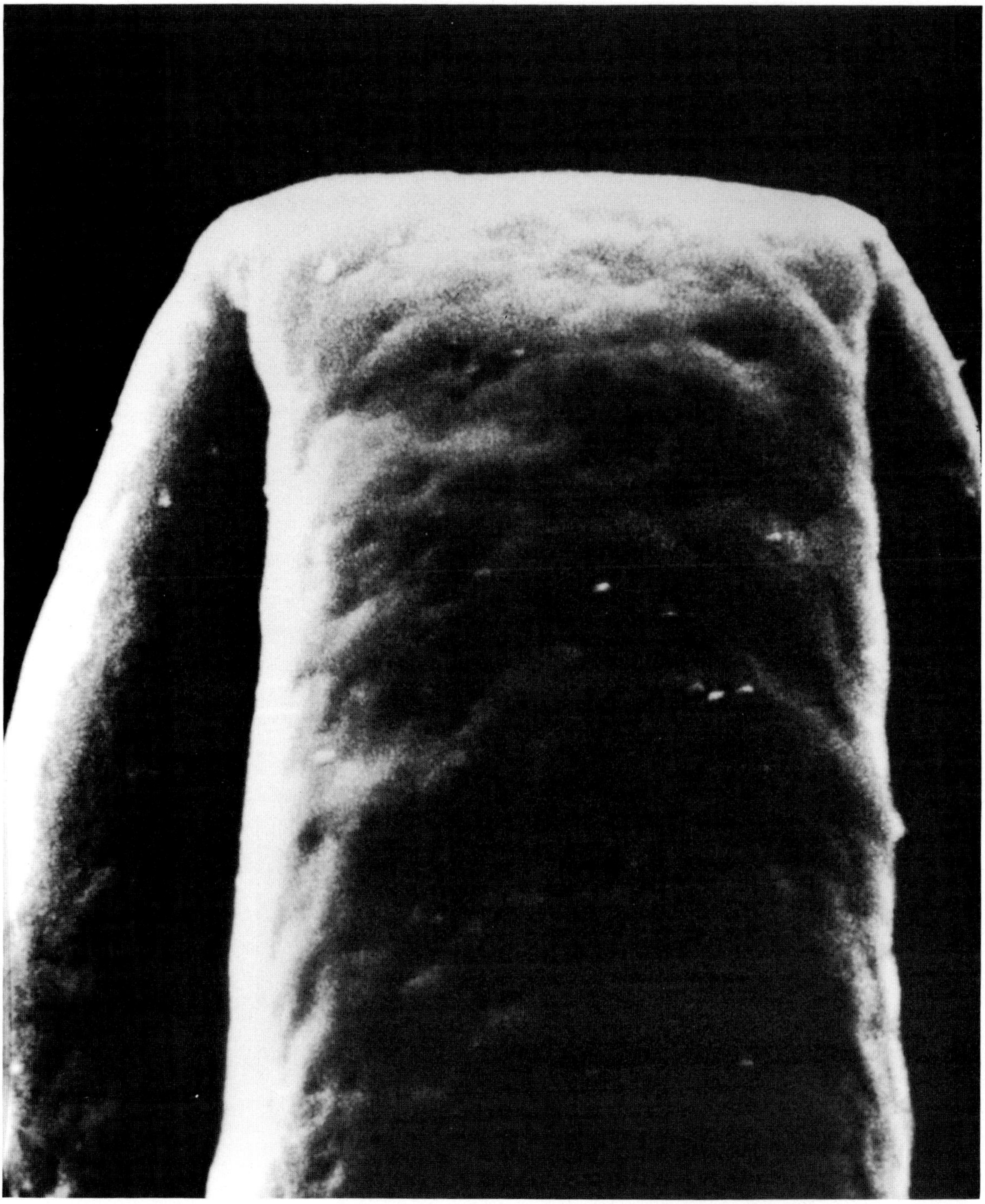

Fig. 2.40 Pollen morphology (18).
Surface of pollen from *Clethra barbinervis* Sieb et Zucc. [**Clethraceae**] (×12500).

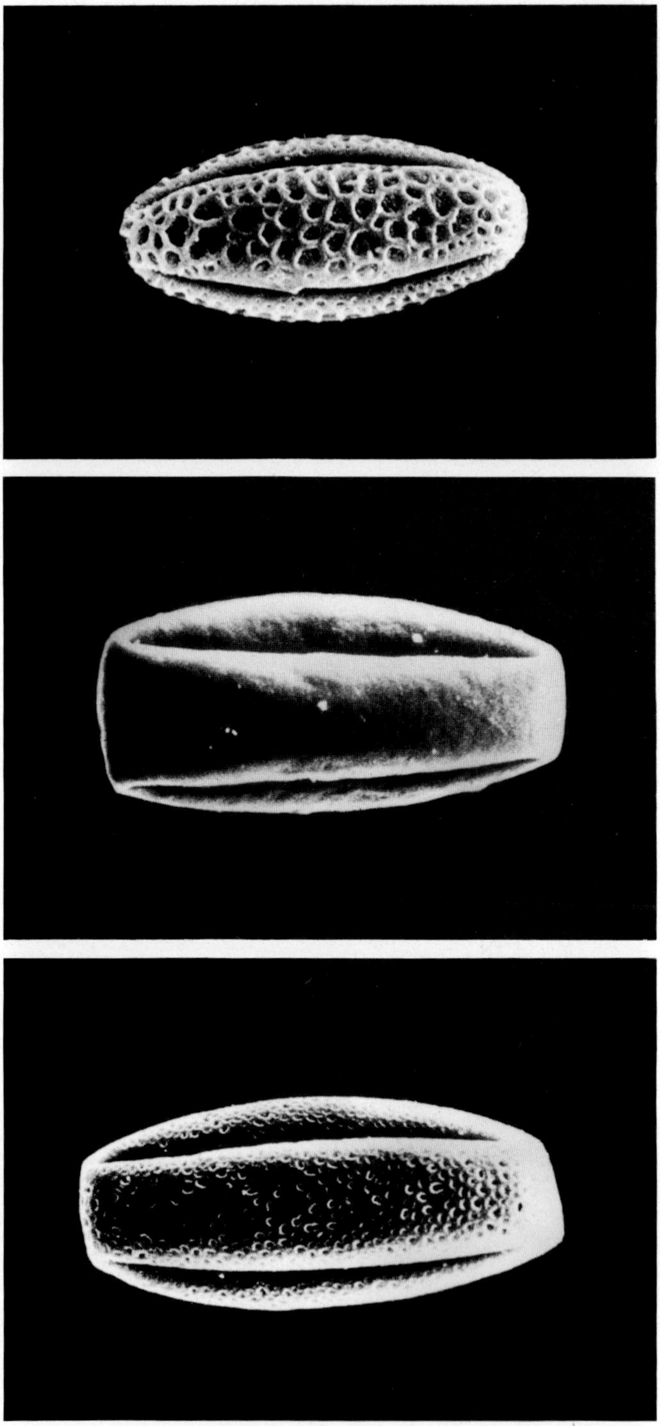

Fig. 2.41 Pollen morphology (19).
Top: Willow (*Salix Bakko* Kimura) [**Salicaceae**] (×1700).
Middle: *Clethra barbinervis* Sieb. et Zucc. [**Clethraceae**] (×1700).
Bottom: *Shortia soldanelloides* Makino [**Diapensiaceae**] (×1700).

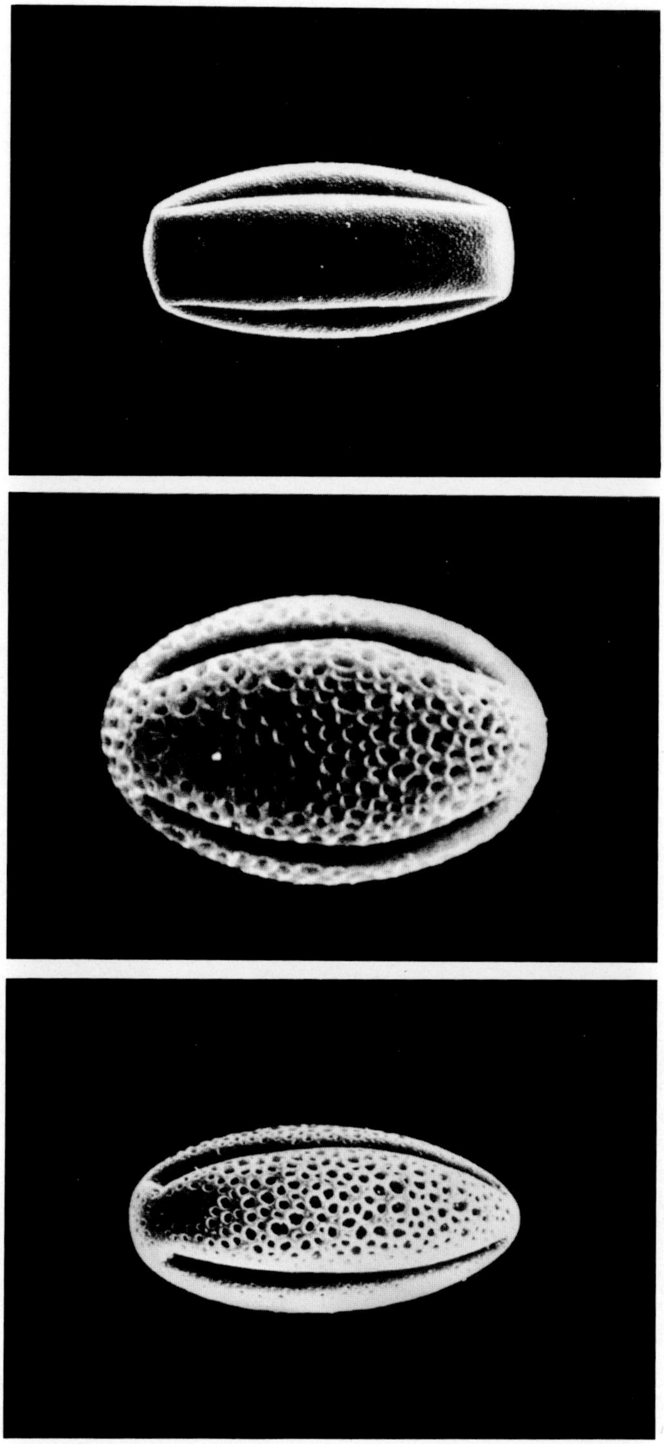

Fig. 2.42 Pollen morphology (20).
Top: Storax (*Styrax japonica* Sieb. et Zucc.) [**Styracaceae**] (×800).
Middle: Witch-Hazel (*Hamamelis japonica* Sieb. et Zucc.) [**Hamamelidaceae**] (×2200).
Bottom: *Adoxa moschatellina* L. [**Adoxaceae**] (×1200).

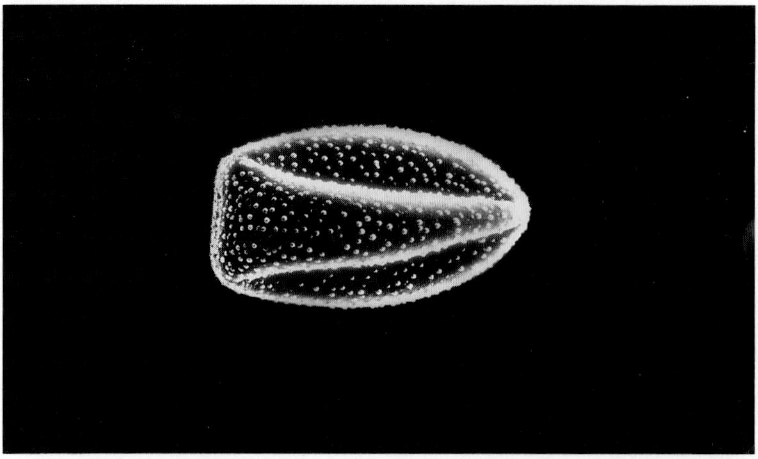

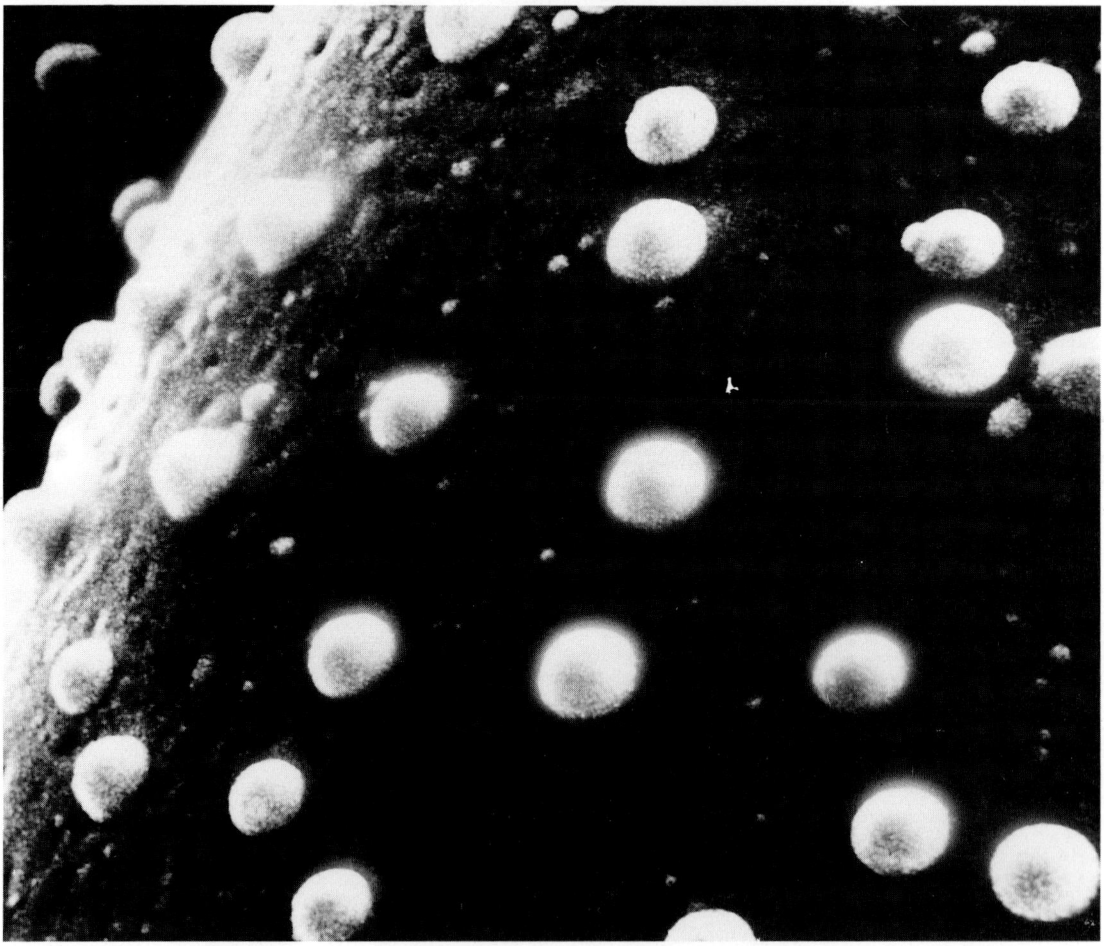

Fig. 2.43 Pollen morphology (21).
Top: Mother-of-Thousands (*Saxifraga stolonifera* Meerb.) [**Saxifragaceae**] (×1300).
Bottom: Detail of ornamentation (×24400).

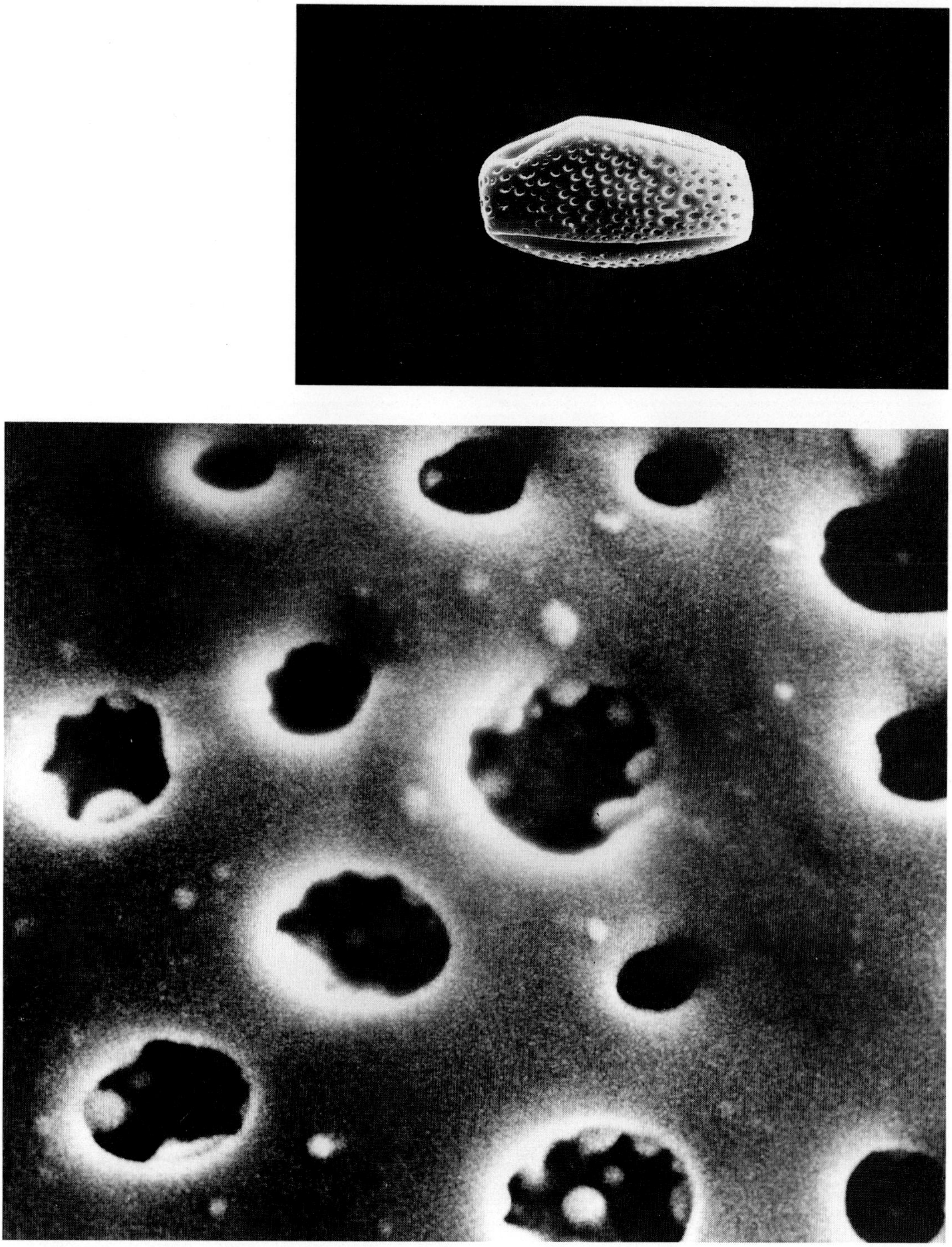

Fig. 2.44 Pollen morphology (22).
Top: Kirengeshoma polmata Yatabe [**Saxifragaceae**] ($\times 1625$).
Bottom: Detail of ornamentation ($\times 24400$).

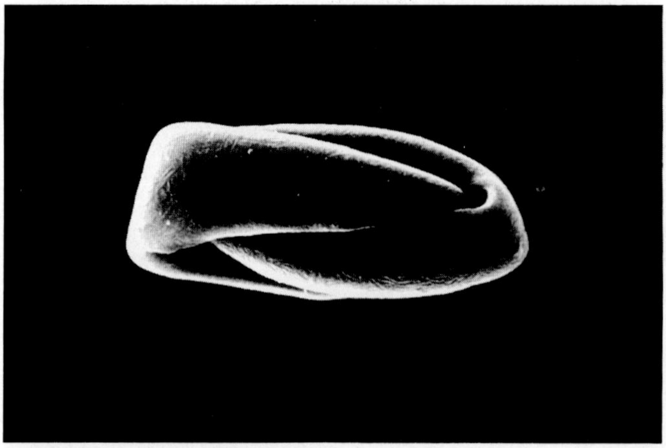

Fig. 2.45 Pollen morphology (23).
Top: Kalanchoë (*Kalanchoe blossfeldiana* Poelln.) **[Crassulaceae]** (×1200).
Bottom: Detail of ornamentation (×22000).

Fig. 2.46 Pollen morphology (24).
Top: Kagetsunishiki (*Crassula portulacea* Lam.) **[Crassulaceae]**. (×1950).
Bottom: Detail of ornamentation (×16250).

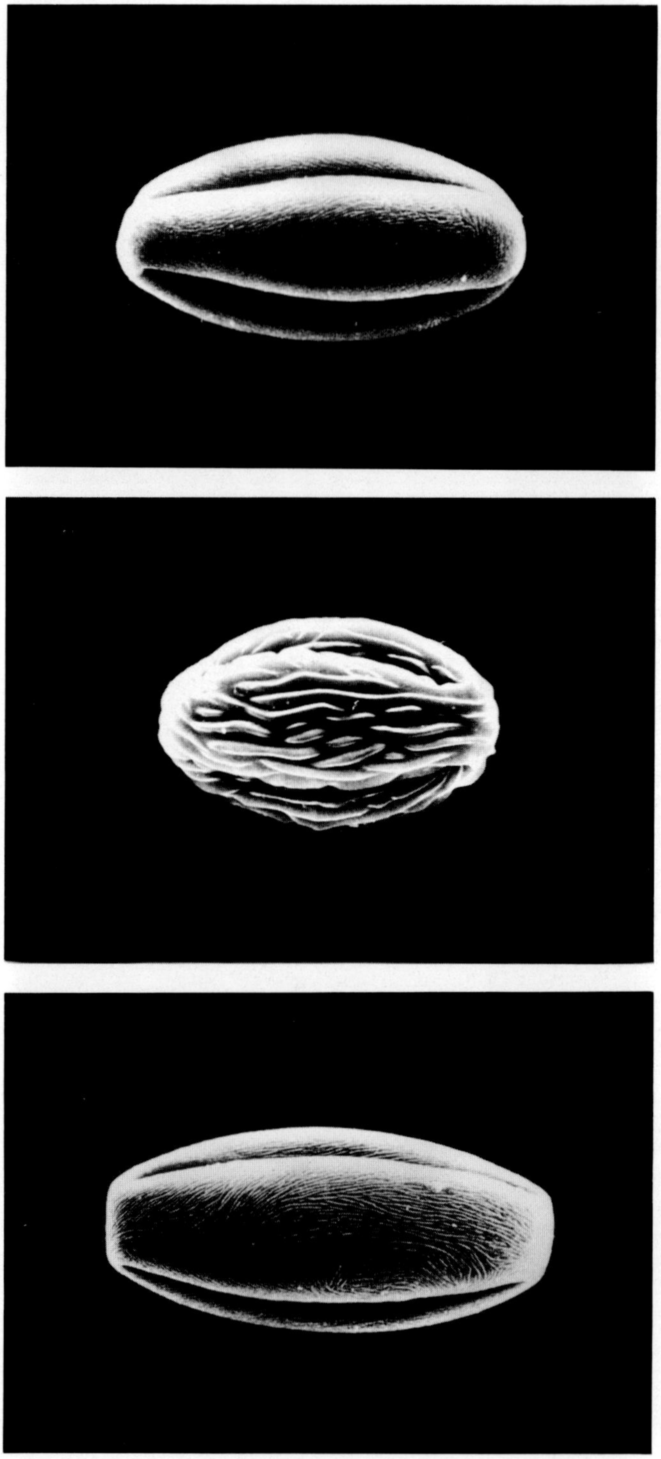

Fig. 2.47 Pollen morphology (25).
Top: Hall's Crab (*Malus Halliana* Koehne) **[Rosaceae]** (×1200).
Middle: Strawberry (*Fragaria chiloensis* Duch. var. *ananassa* Bailey) **[Rosaceae]** (×1600).
Bottom: Toringo Crab (Malus Toringo Sieb.) **[Rosaceae]** (×1100).

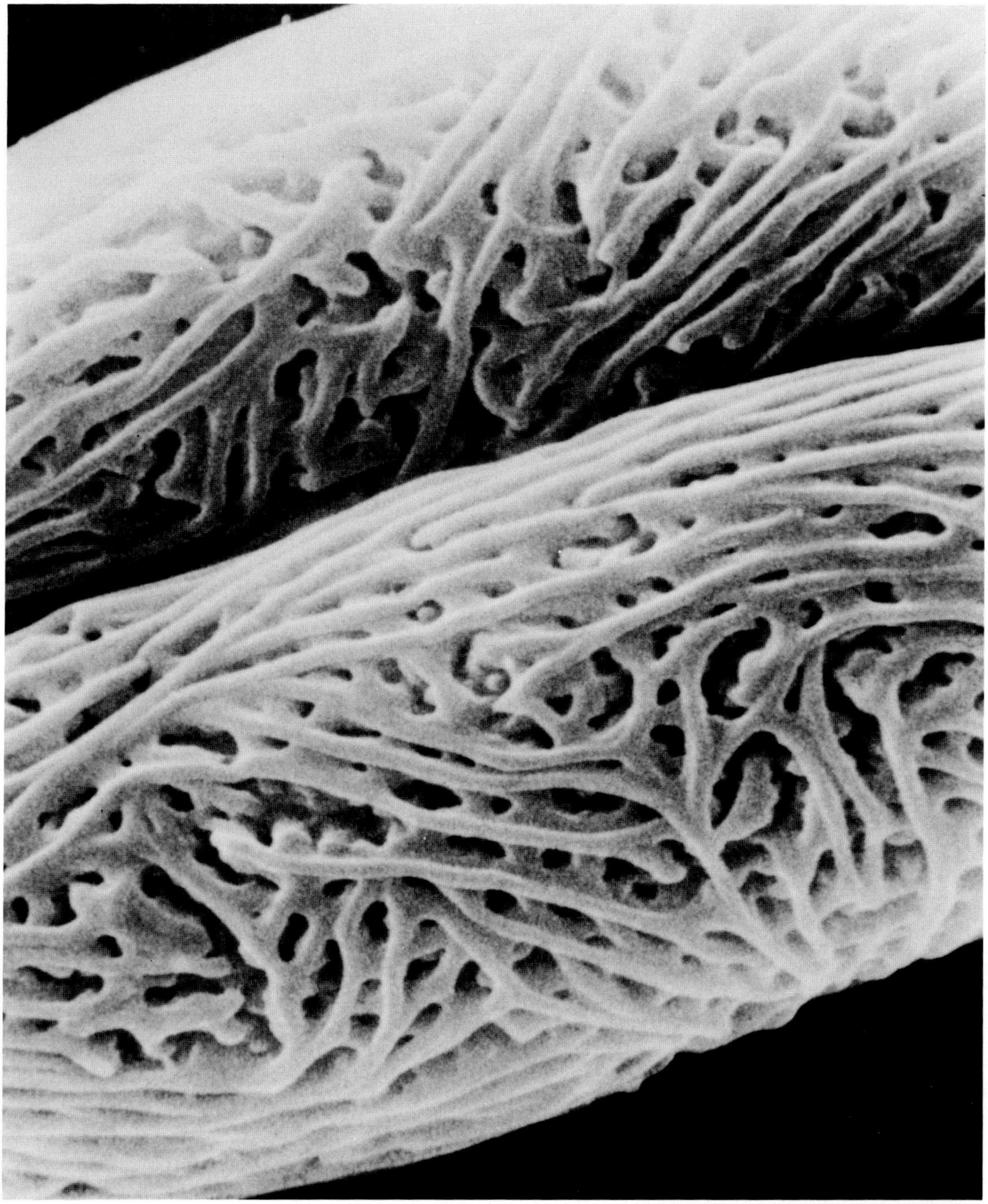

Fig. 2.48 Pollen morphology (26).
Surface of pollen from Japanese Apricot (*Prunus Mume* Sieb. et Zucc.) **[Rosaceae]** ($\times$18800).

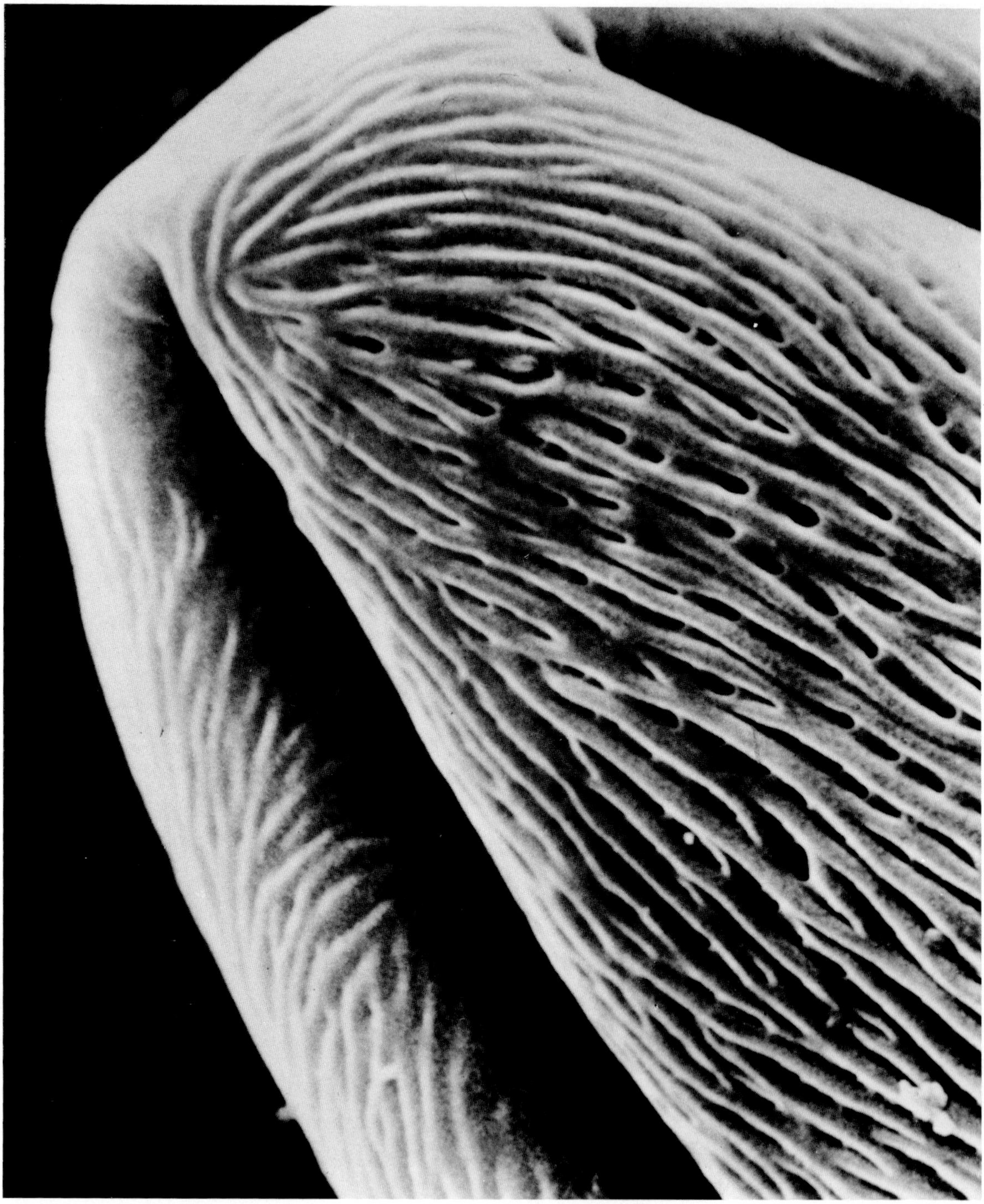

Fig. 2.49 Pollen morphology (27).
Surface of pollen from Peach (*Prunus Persica* Batsch) **[Rosaceae]** (×11000).

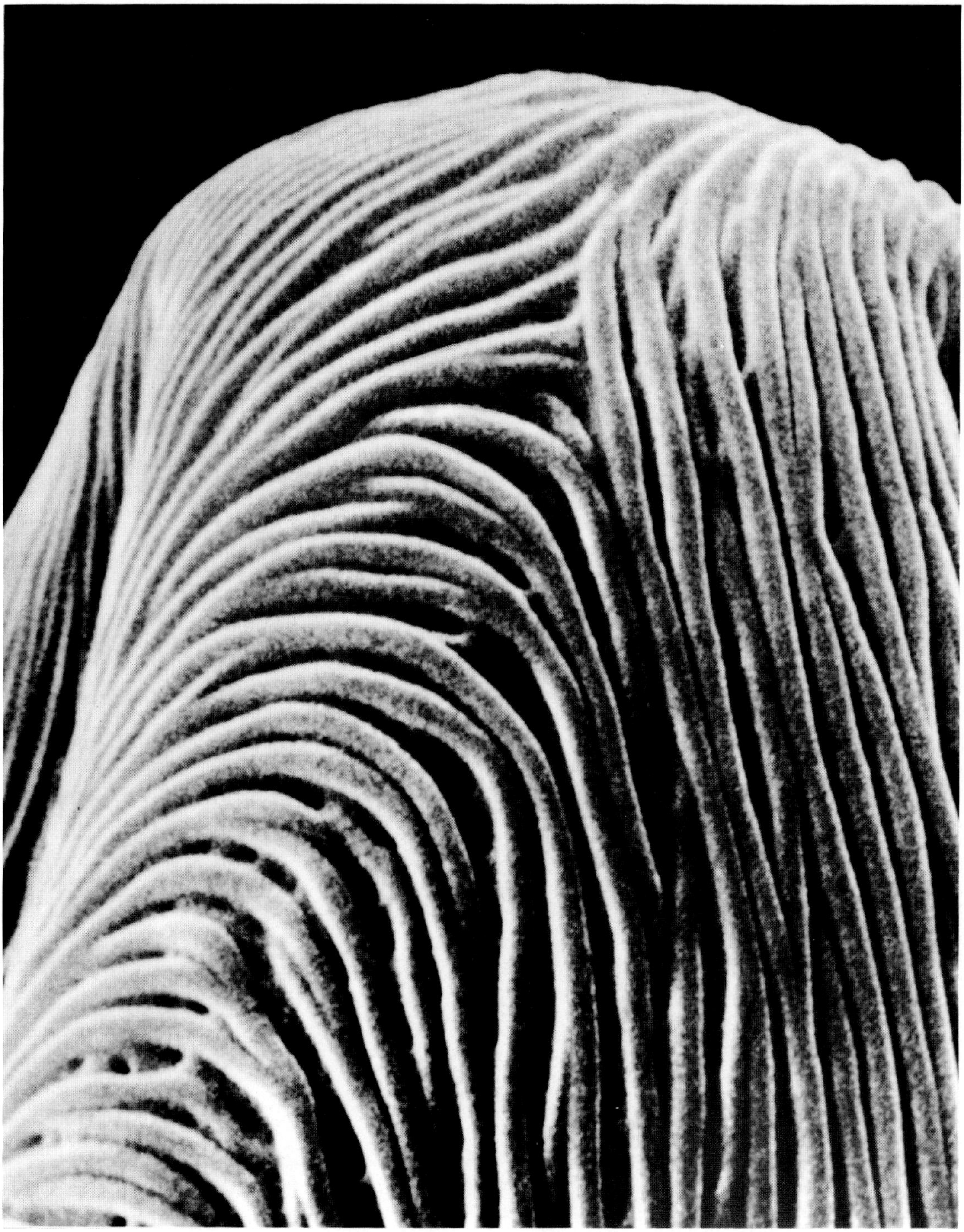

Fig. 2.50 Pollen morphology (28).
Surface of pollen from Japanese Pear (*Pyrus pyrifolia* Nakai) **[Rosaceae]** (×17000).

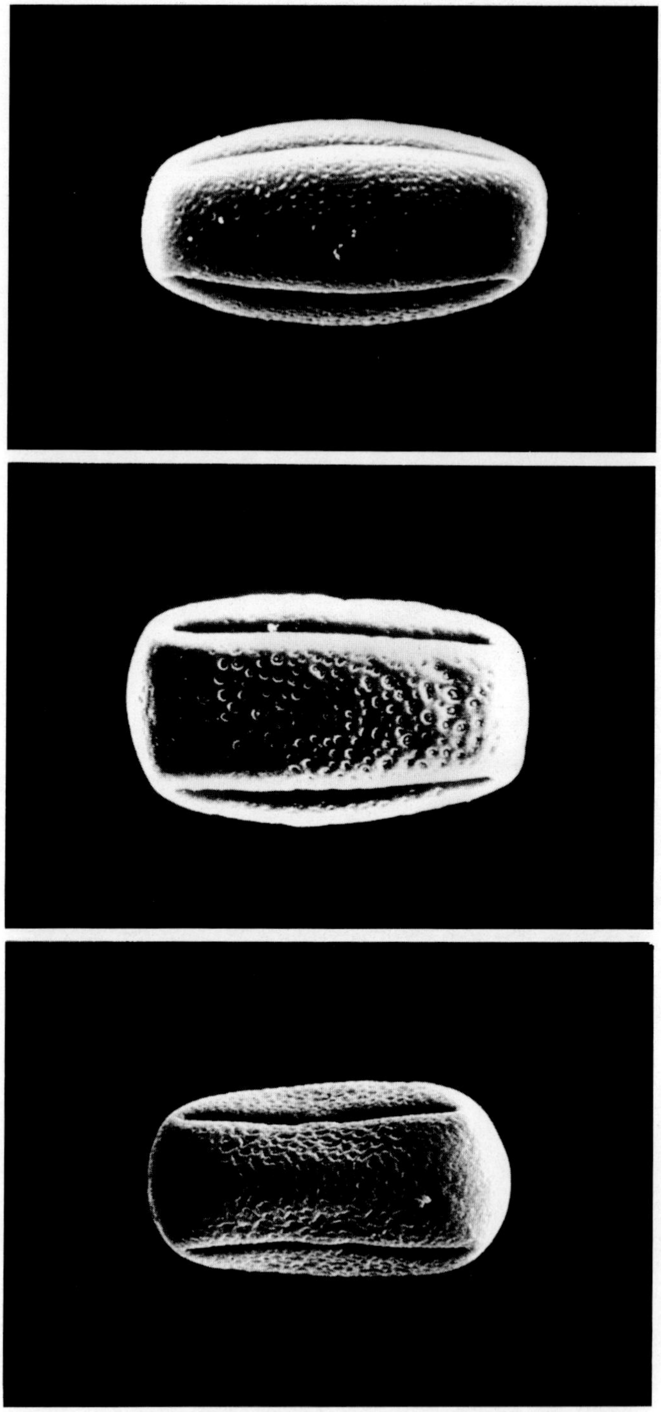

Fig. 2.51 Pollen morphology (29).
Top: Lace Shrub (*Stephanandra incisa* Zabel) [**Rosaceae**] (×1700).
Middle: Japanese Wistaria (*Wisteria floribunda* DC.) [**Leguminosae**] (×1200).
Bottom: Sweet Pea (*Lathyrus odoratus* L.) [**Leguminosae**] (×800).

Fig. 2.52 Pollen morphology (30).
Surface of pollen from Milk-Vetch (*Astragalus sinicus* L.) **[Leguminosae]** (×18800).

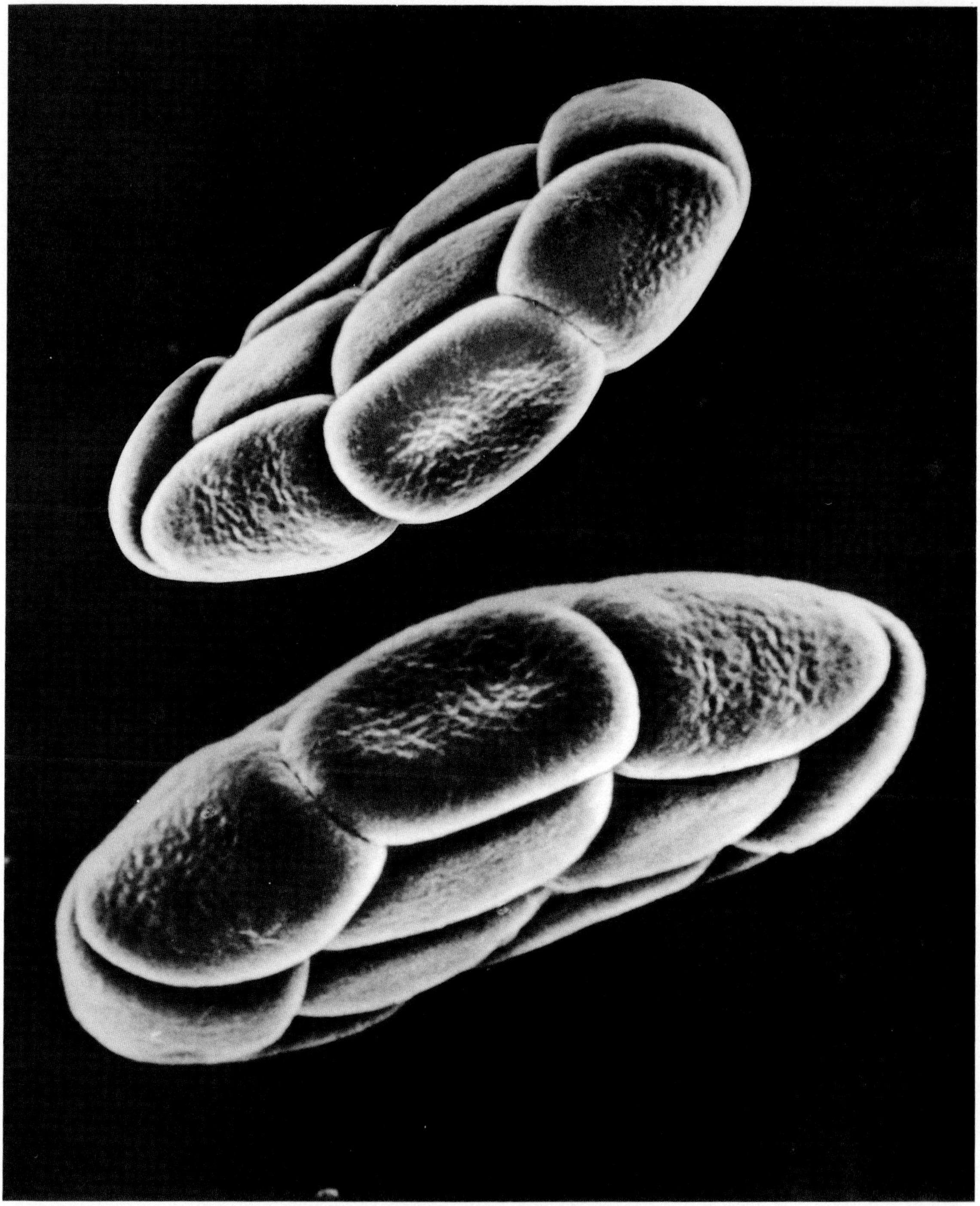

Fig. 2.53 Pollen morphology (31).
Silk Flower (*Albizzia julibrissin* Durazz.) [**Leguminosae**] (×1800).

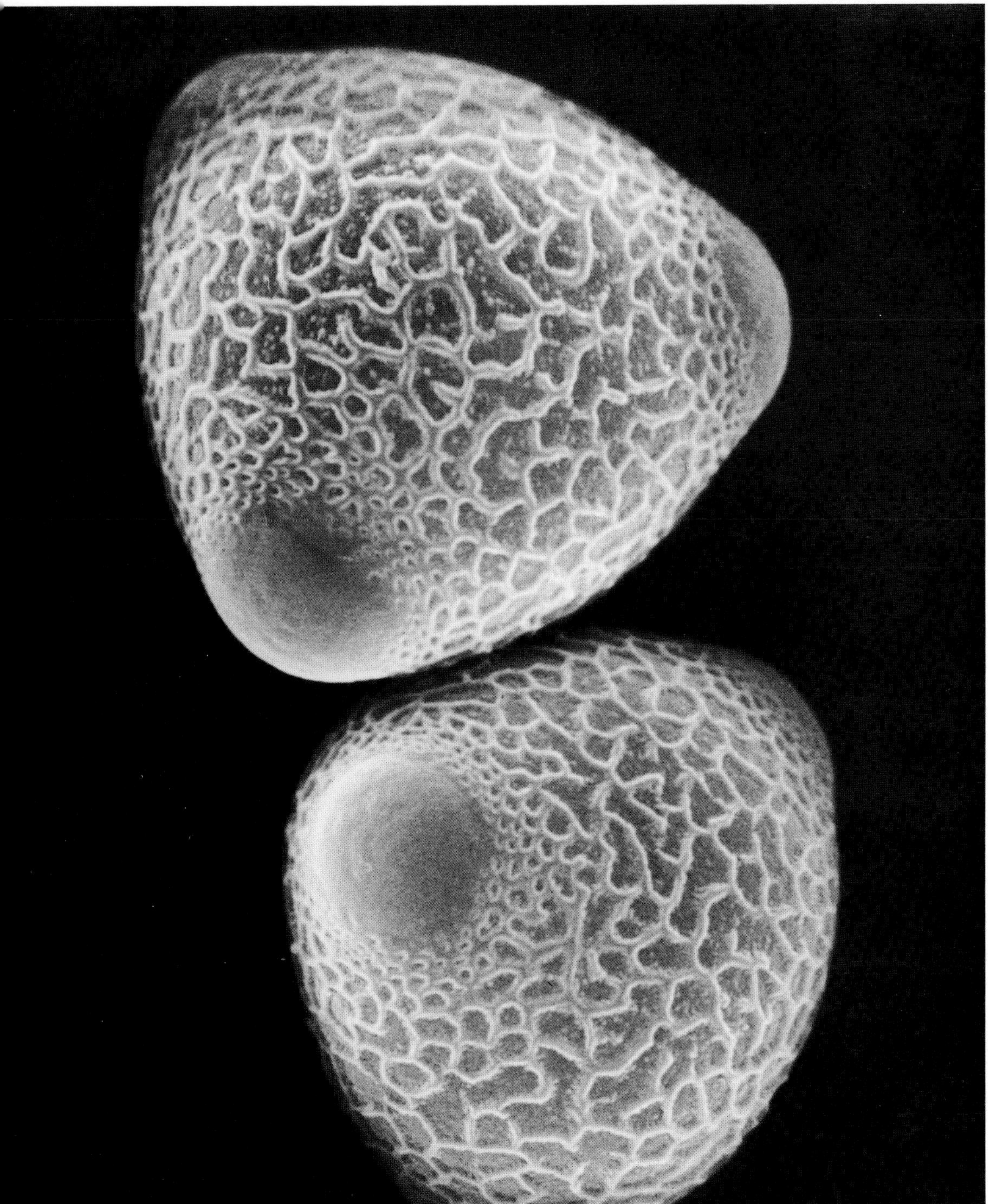

Fig. 2.54 Pollen morphology (32).
Cockspur Coral Tree (*Erythrina crista-galli* L.) [**Leguminosae**] (×4400).

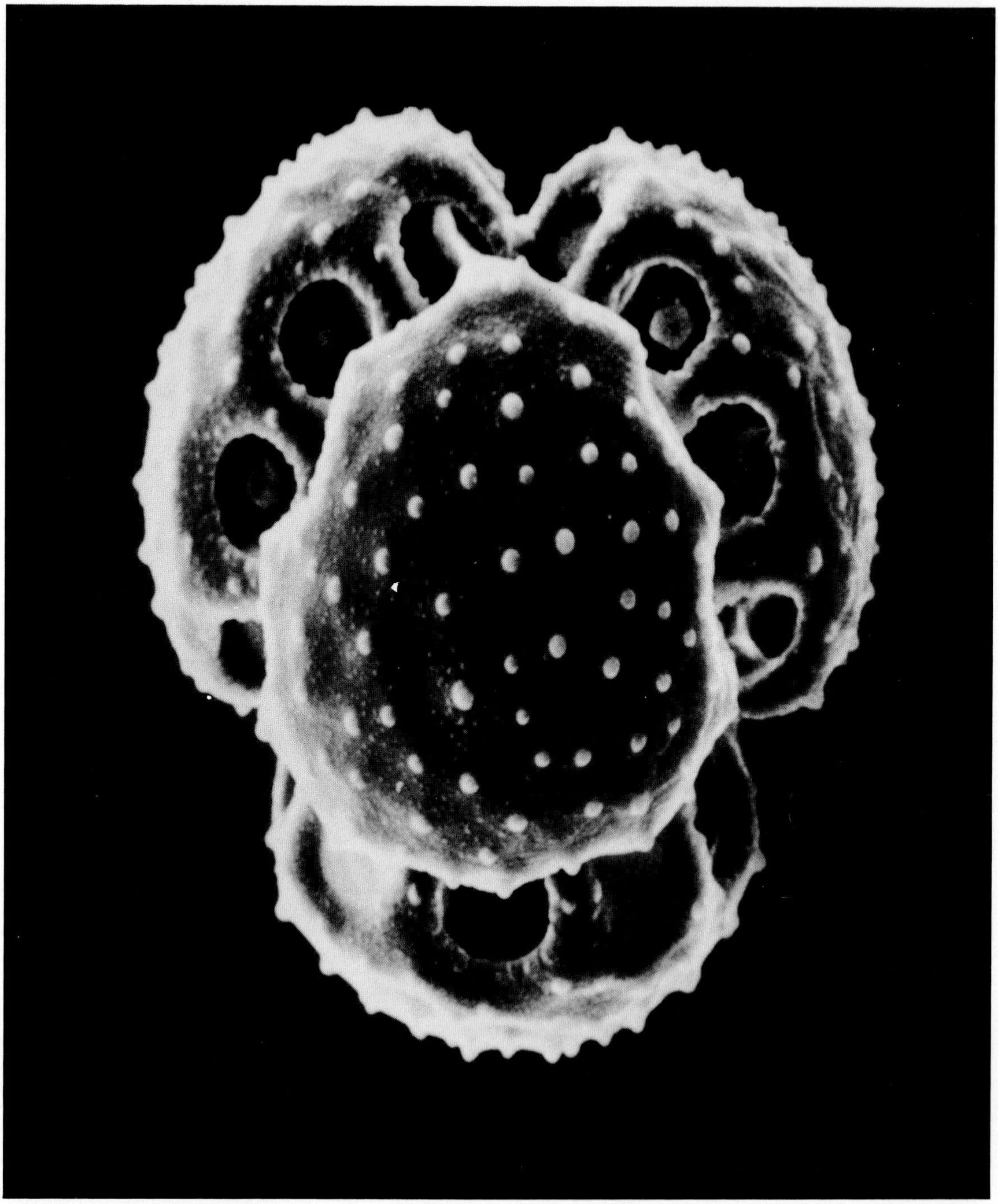

Fig. 2.55 Pollen morphology (33).
Venus's Fly Trap (*Dionaea muscipula* Ellis) [**Droseraceae**] ($\times 1300$).

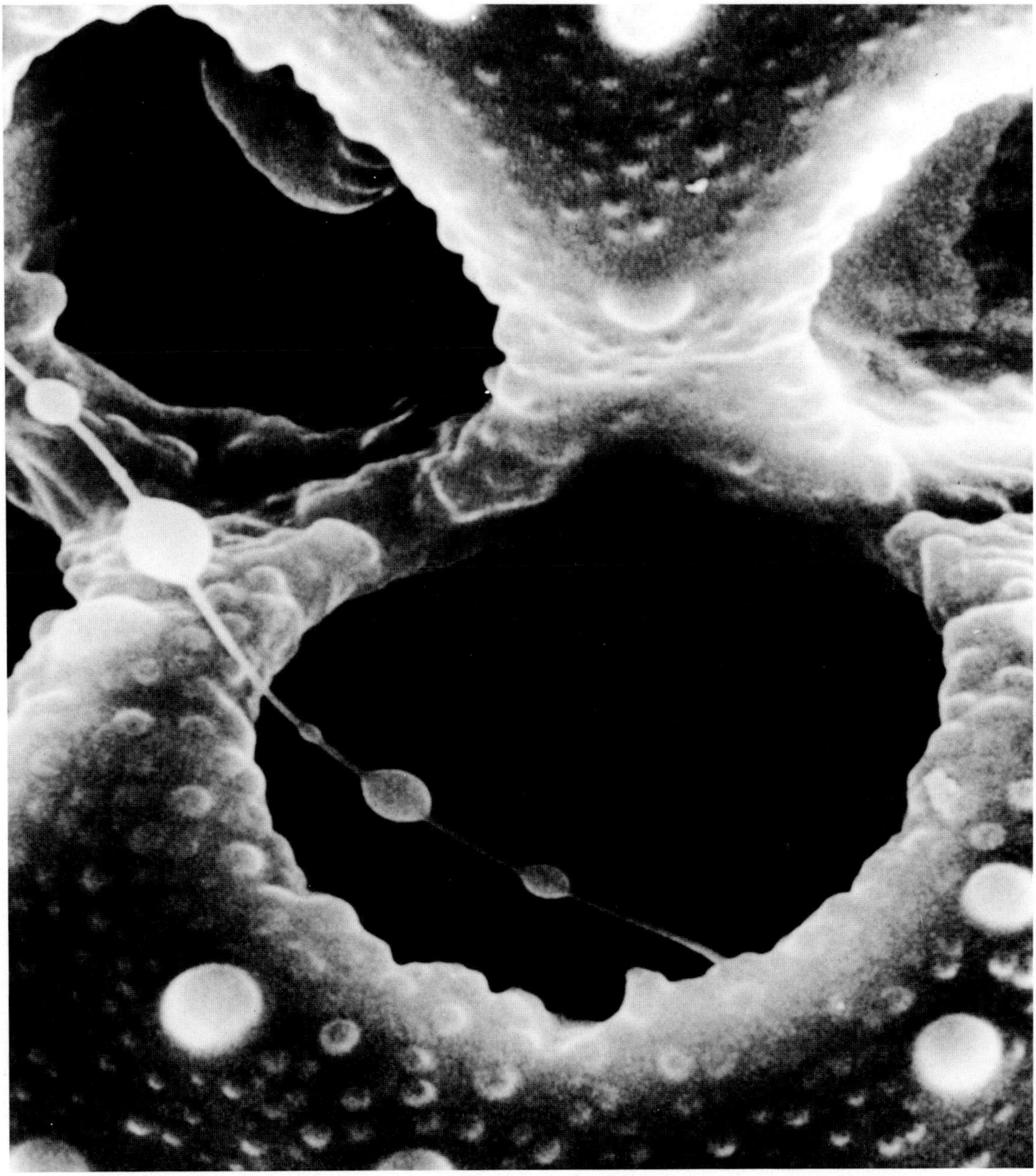

Fig. 2.56 Pollen morphology (34).
Surface of pollen from Venus's Fly Trap (*Dionaea muscipula* Ellis) (×5600).

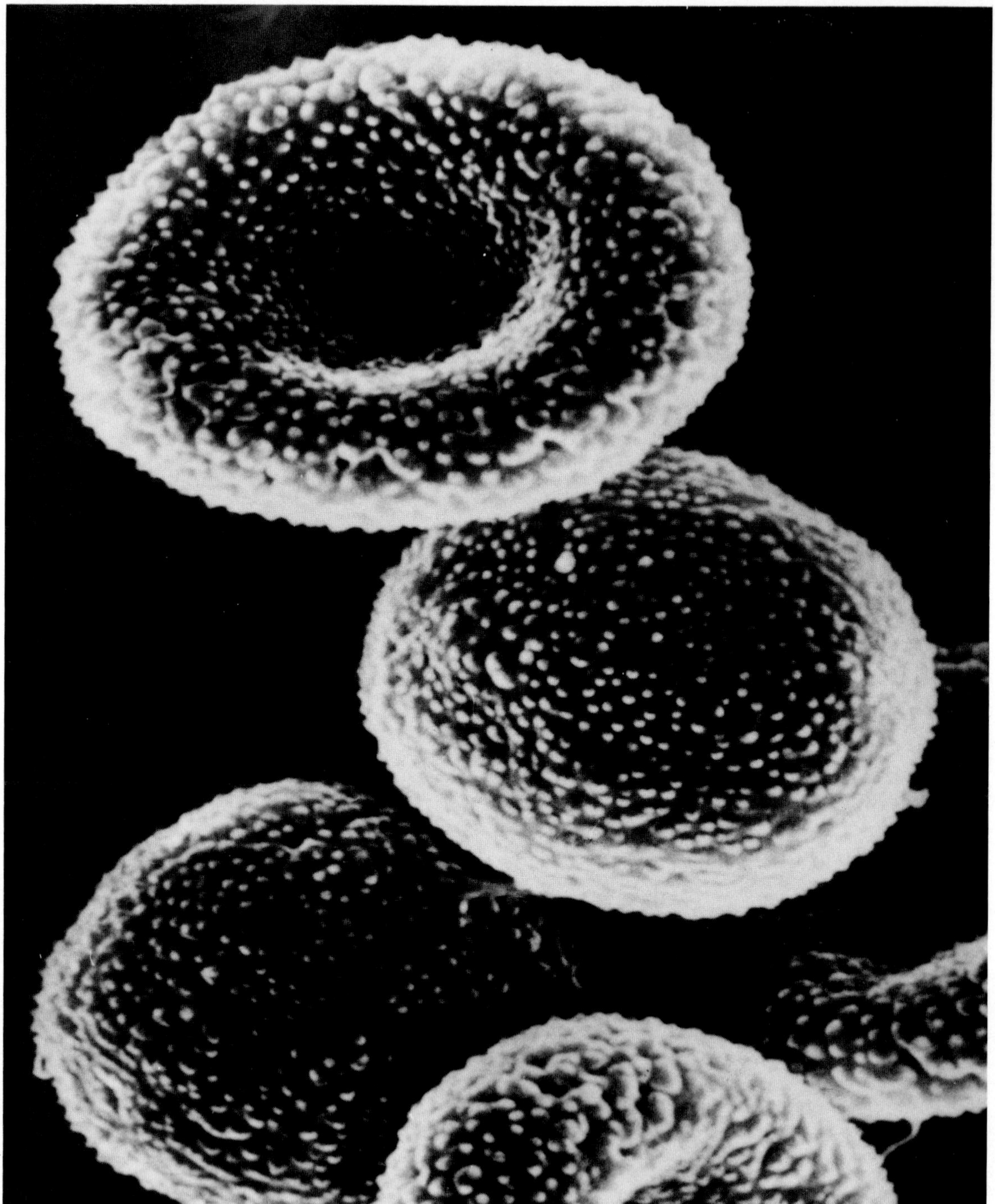

Fig. 2.57 Pollen morphology (35).
Winter Daphne (*Daphne odora* Thunb.) [**Thymelaeaceae**] ($\times 3200$).

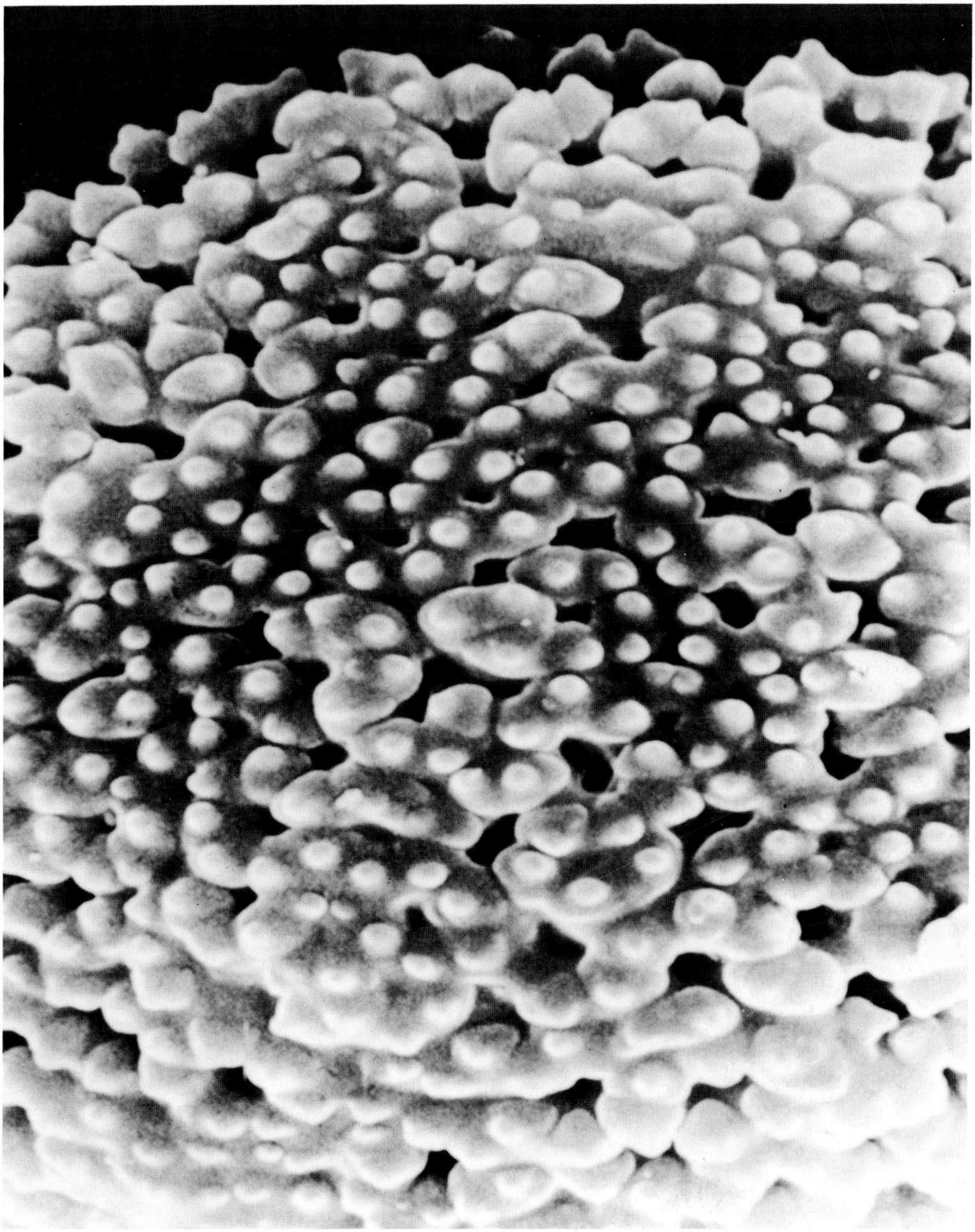

Fig. 2.58 Pollen morphology (36).
Surface of pollen from Winter Daphne (*Daphne odora* Thunb.) ($\times 10000$).

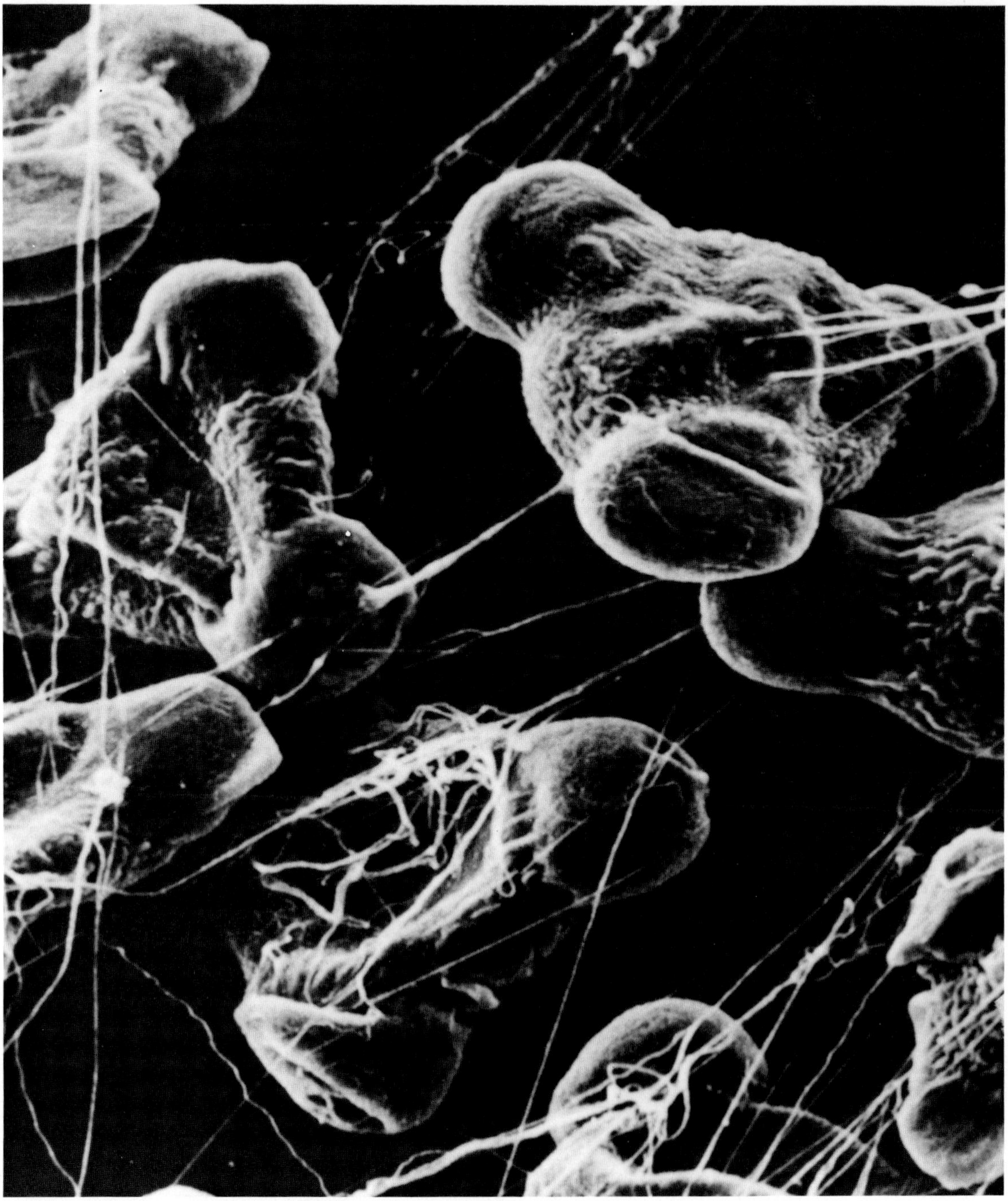

Fig. 2.59 Pollen morphology (37).
Evening Primrose (*Oenothera biennis* L.) [**Onagraceae**] (×2000).
Group of pollen with viscin threads.

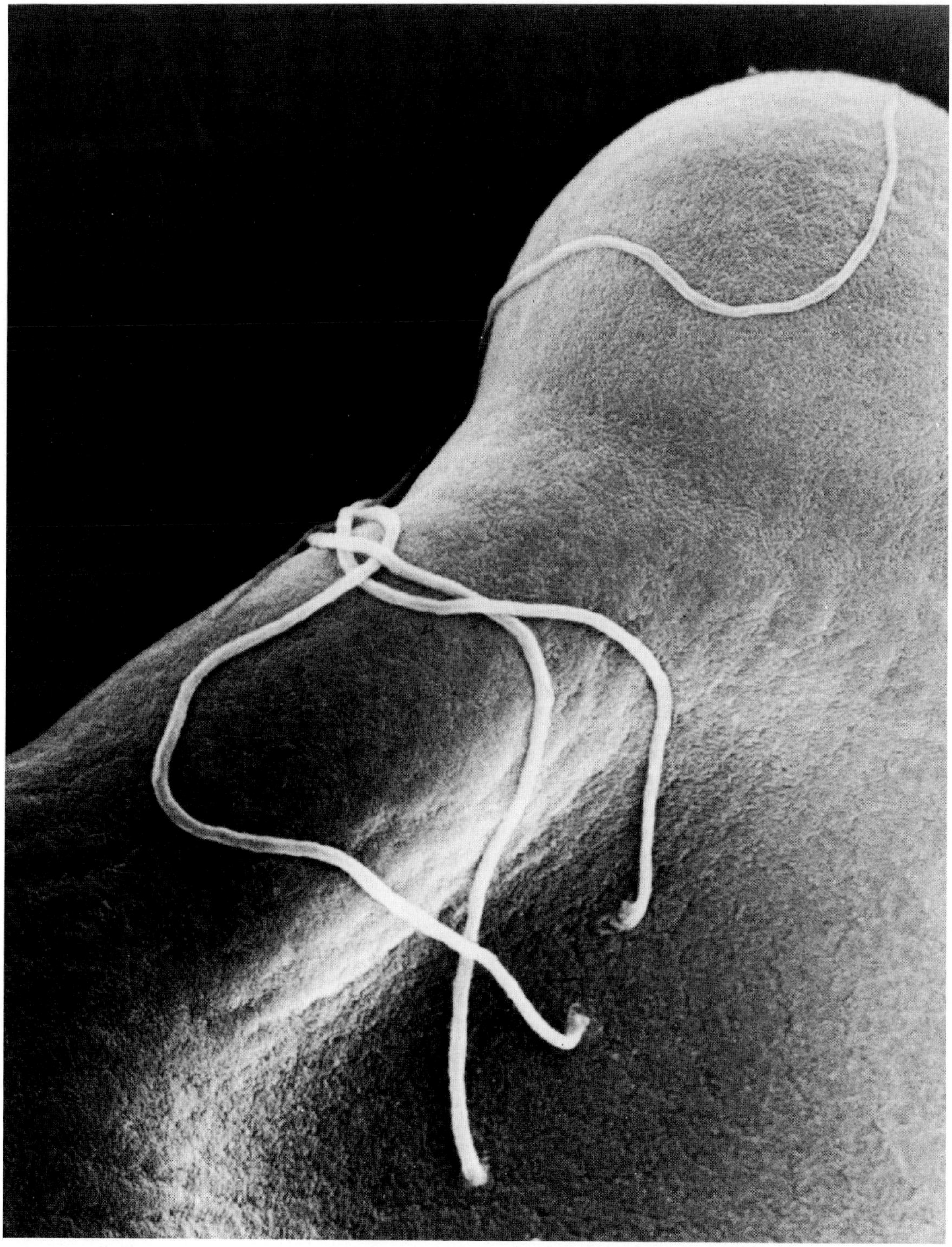

Fig. 2.60 Pollen morphology (38).
Surface of pollen from *Oenothera odorata* Jacq. **[Onagraceae]** (× 4500).
Rope-like structures are viscin threads.

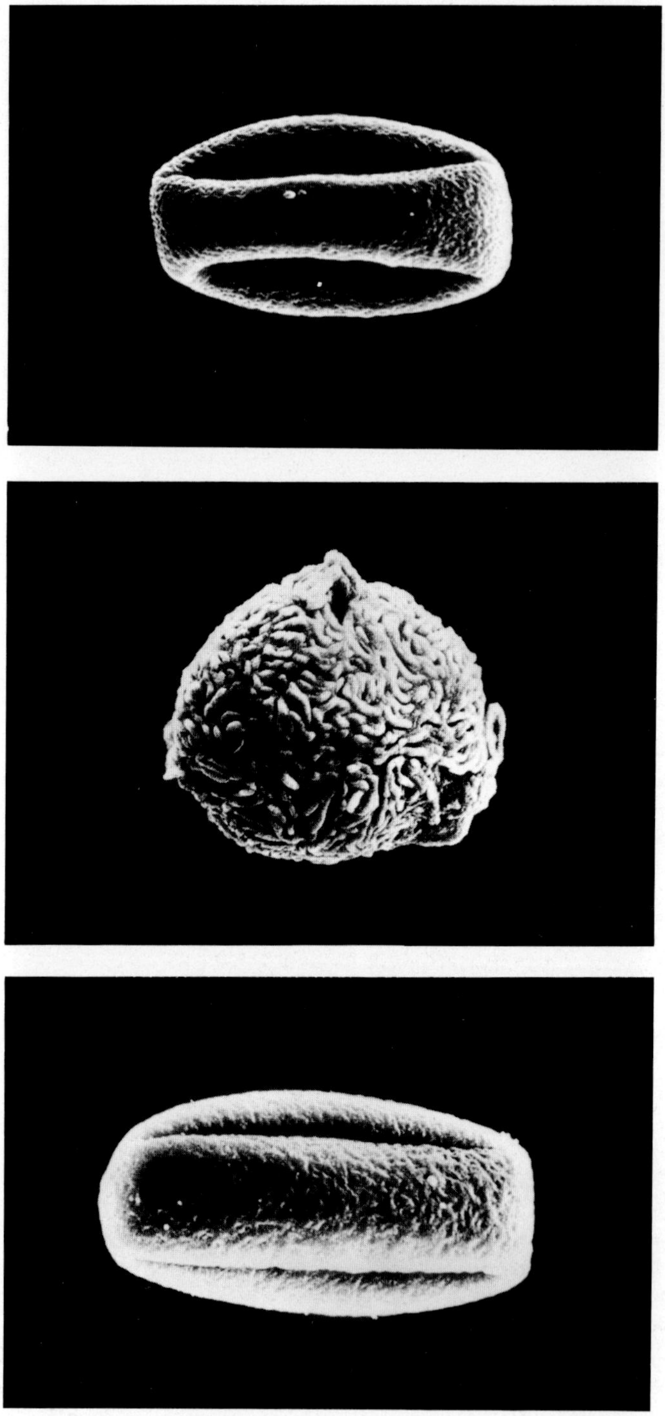

Fig. 2.61 Pollen morphology (39).
Top: Mangrove (*Rhizophora mucronata* Lam.) [**Rhizophoraceae**] (×1700).
Middle: *Alangium platanifolium* Harms var. *macrophyllum* Wangerin [**Alangiaceae**] (×500).
Bottom: Japanese Helwing (*Helwingia japonica* Dietr.) [**Cornaceae**] (×1700).

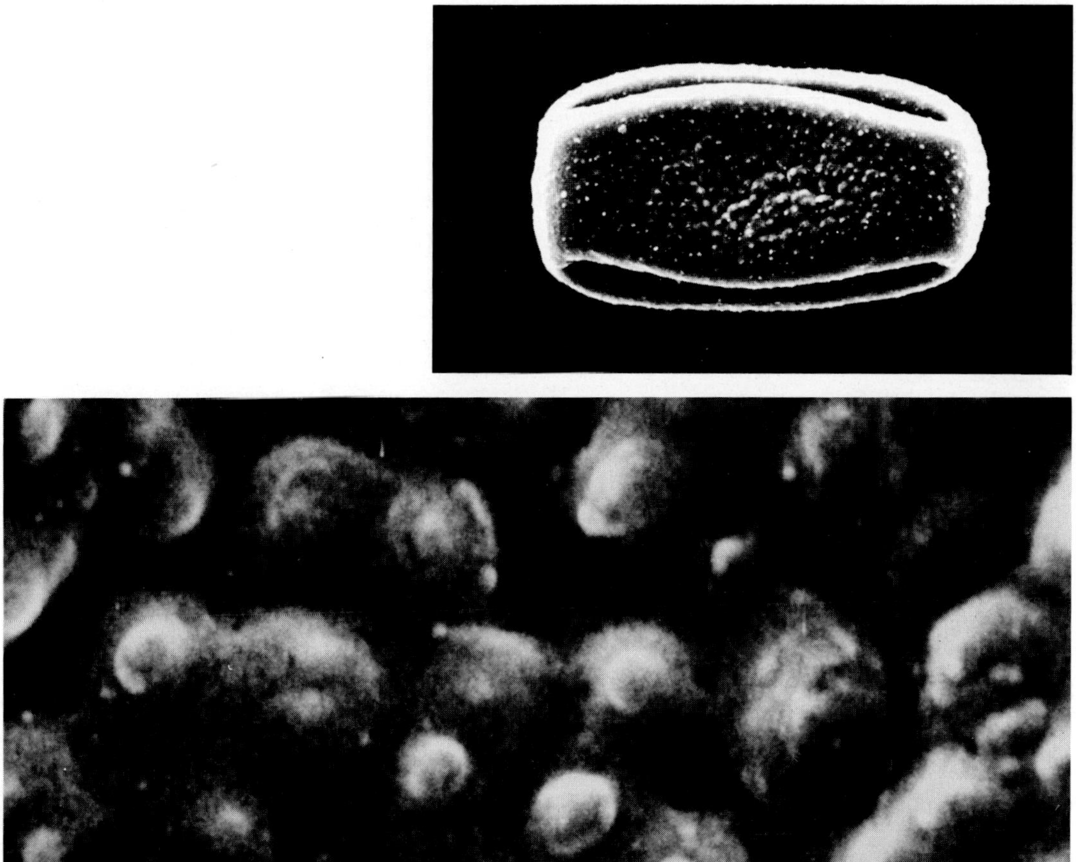

Fig. 2.62 Pollen morphology (40).
Top: Dogwood (*Cornus controversa* Hemsl.) [**Cornaceae**] (×1100).
Bottom: Detail of ornamentation (×15600).

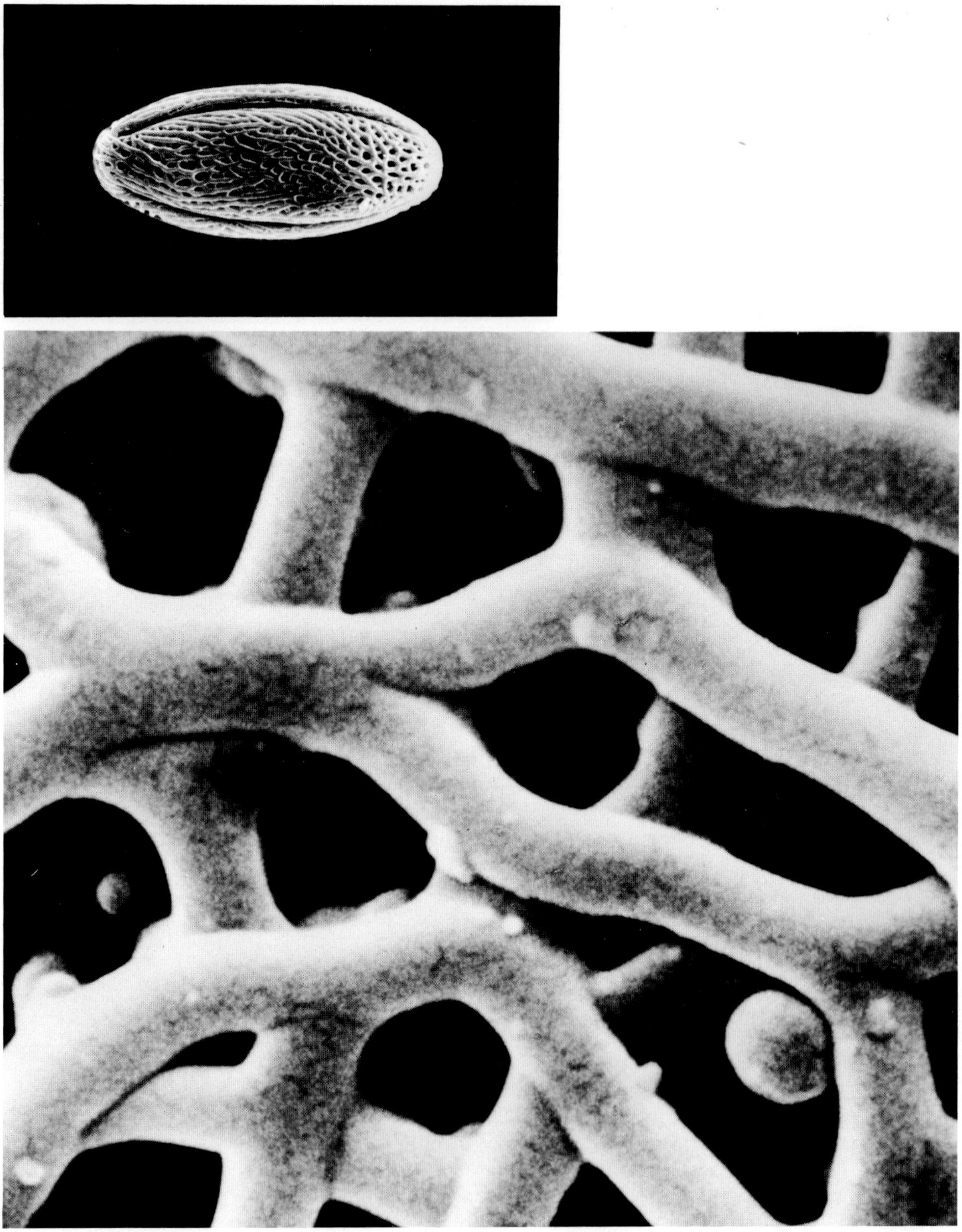

Fig. 2.63 Pollen morphology (41).
Top: Varnish Tree (*Rhus verniciflua* Stockes) [**Anacardiaceae**] (×1200).
Bottom: Detail of ornamentation (×28000).

Fig. 2.64 Pollen morphology (42).
Surace of pollen from Prickly-ash (*Zanthoxylum piperitum* DC.) [**Rutaceae**] (×20000).

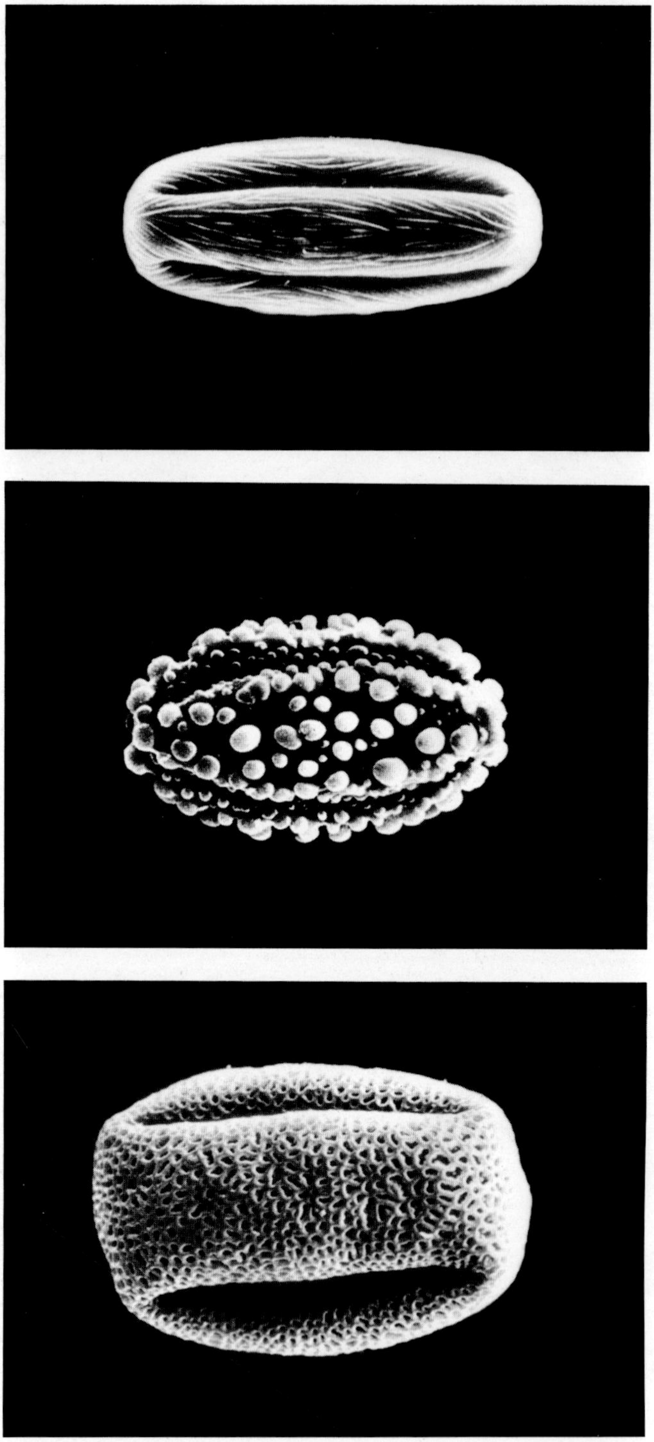

Fig. 2.65 Pollen morphology (43).
Top: *Orixa japonica* Thunb. **[Rutaceae]** (×1200).
Middle: Holly (*Ilex macropoda* Miq.) **[Aquifoliaceae]** (×1200).
Bottom: Creeping Lady's-Sorrel (*Oxalis fontana* Bunge) **[Oxalidaceae]** (×1200).

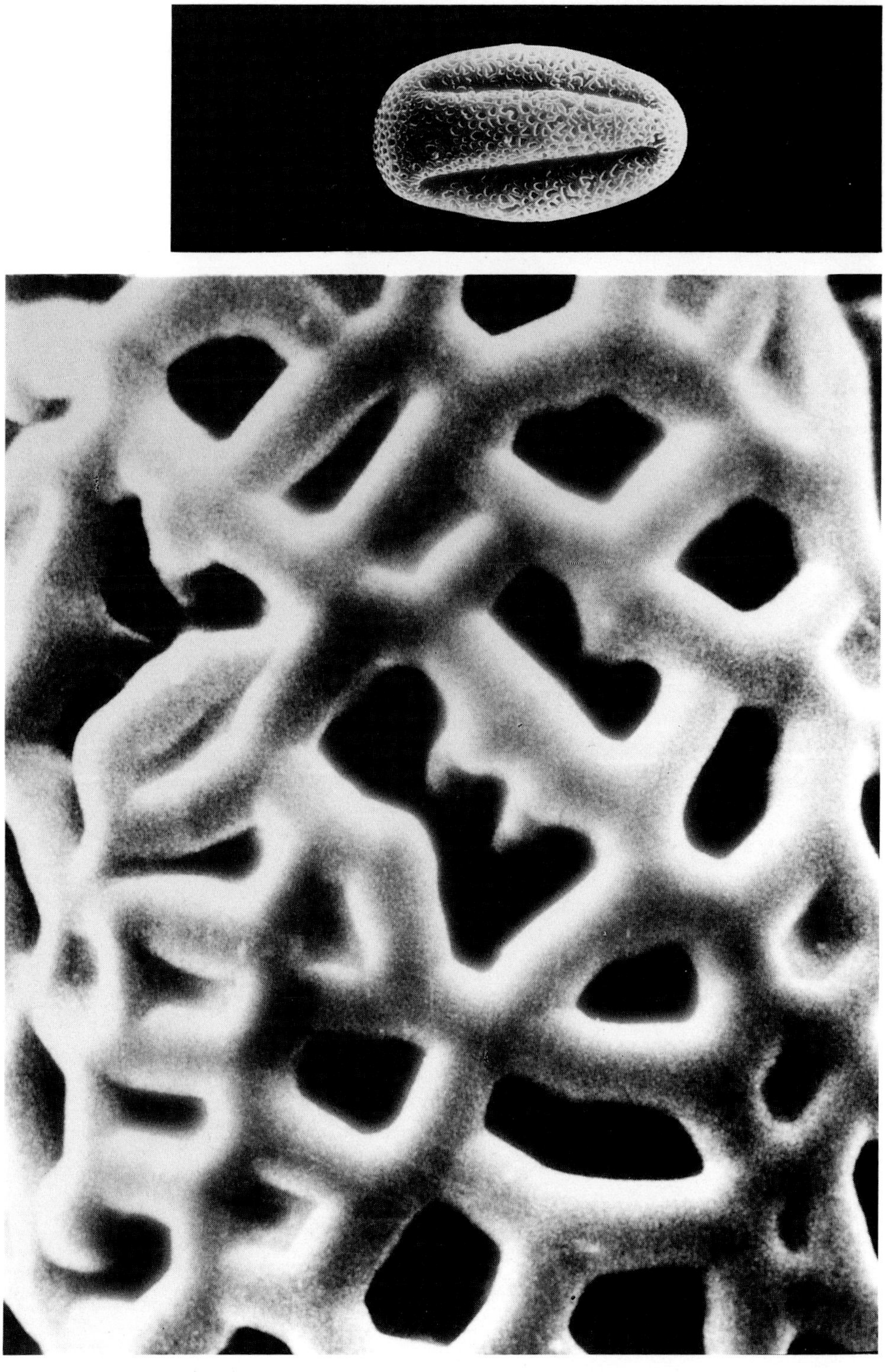

Fig. 2.66 Pollen morphology (44).
Top: Trifoliate-orange (*Poncirus trifoliata* Rafin.) **[Rutaceae]** ($\times$1300).
Bottom: Detail of ornamentation ($\times$24400).

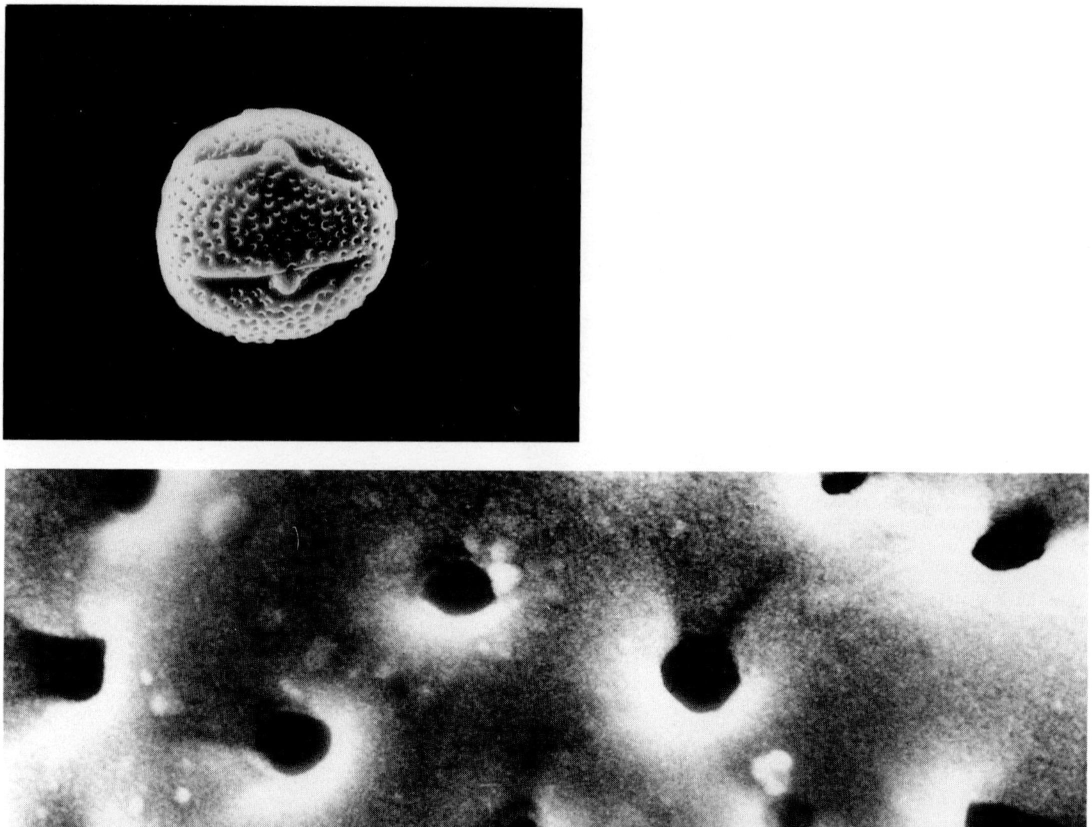

Fig. 2.67 Pollen morphology (45).
Top: Satsuma orange (*Citrus Unshiu* Marcov.) **[Rutaceae]** (×1300).
Bottom: Detail of ornamentation (×24400).

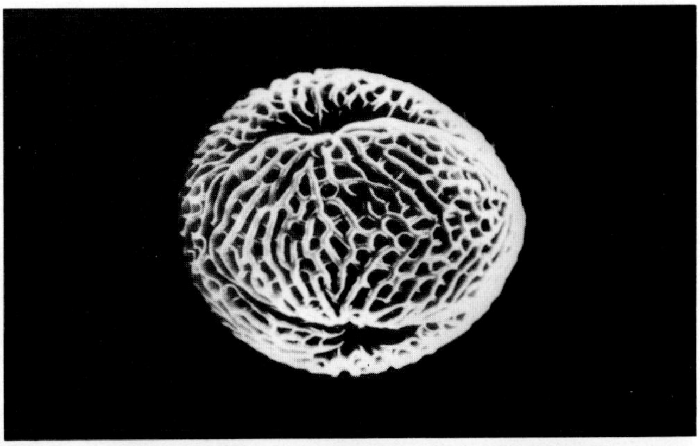

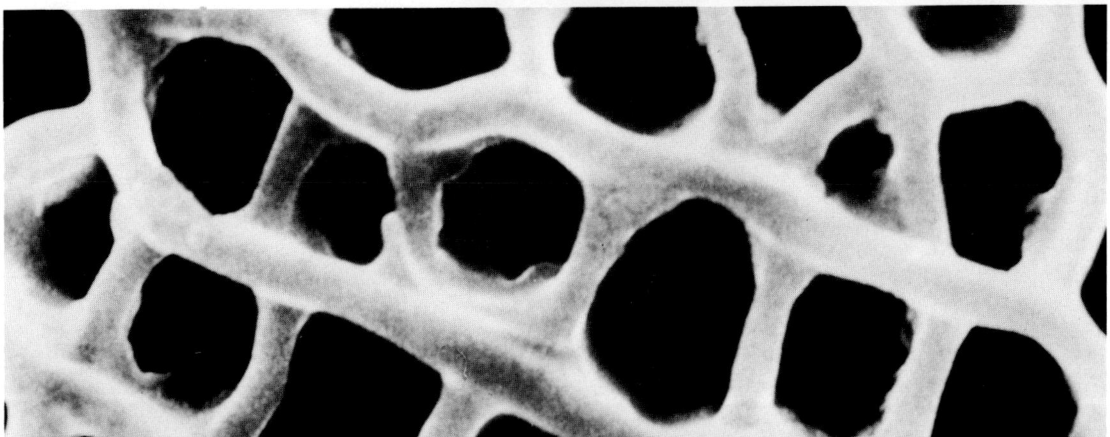

Fig. 2.68 Pollen morphology (46).
Top: Blue-flowering garden perennial (*Pelargonium domesticum* hybrid) **[Geraniaceae]** (×400).
Bottom: Detail of ornamentation (×1500).

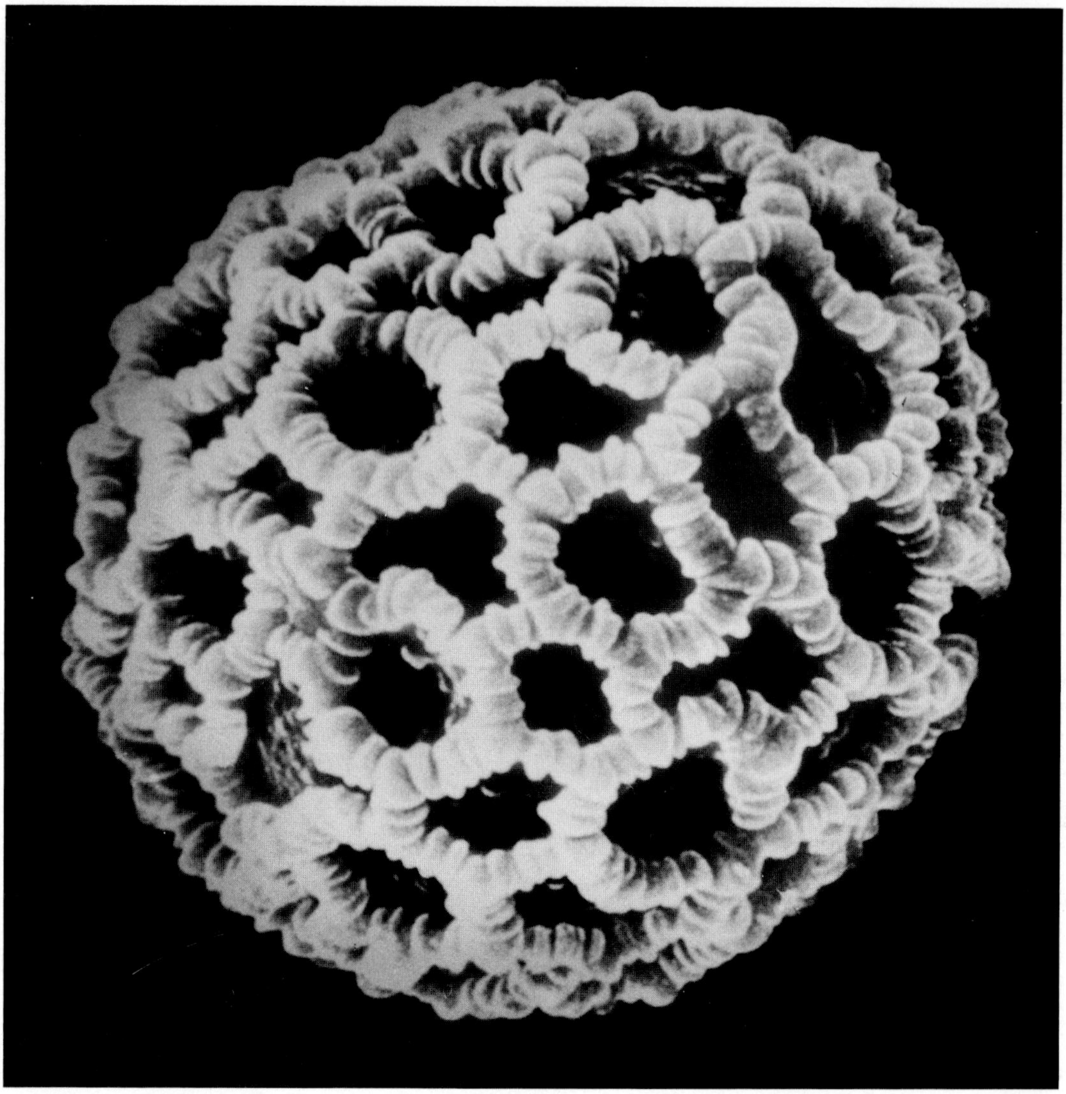

Fig. 2.69 Pollen morphology (47).
Japanese Spurge (*Pachysandra terminalis* Sieb. et Zucc.) **[Buxaceae]** (×1100).

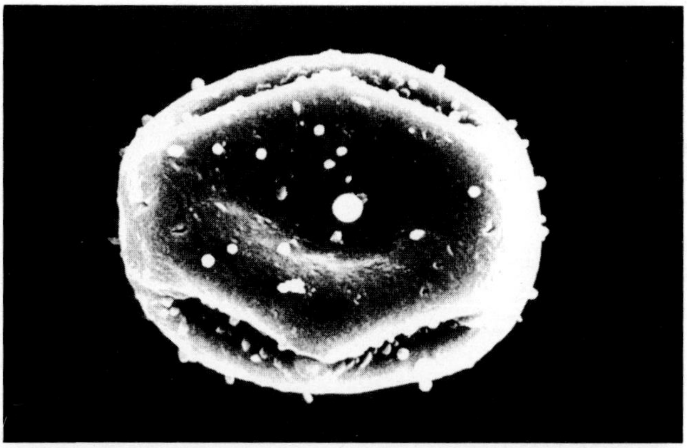

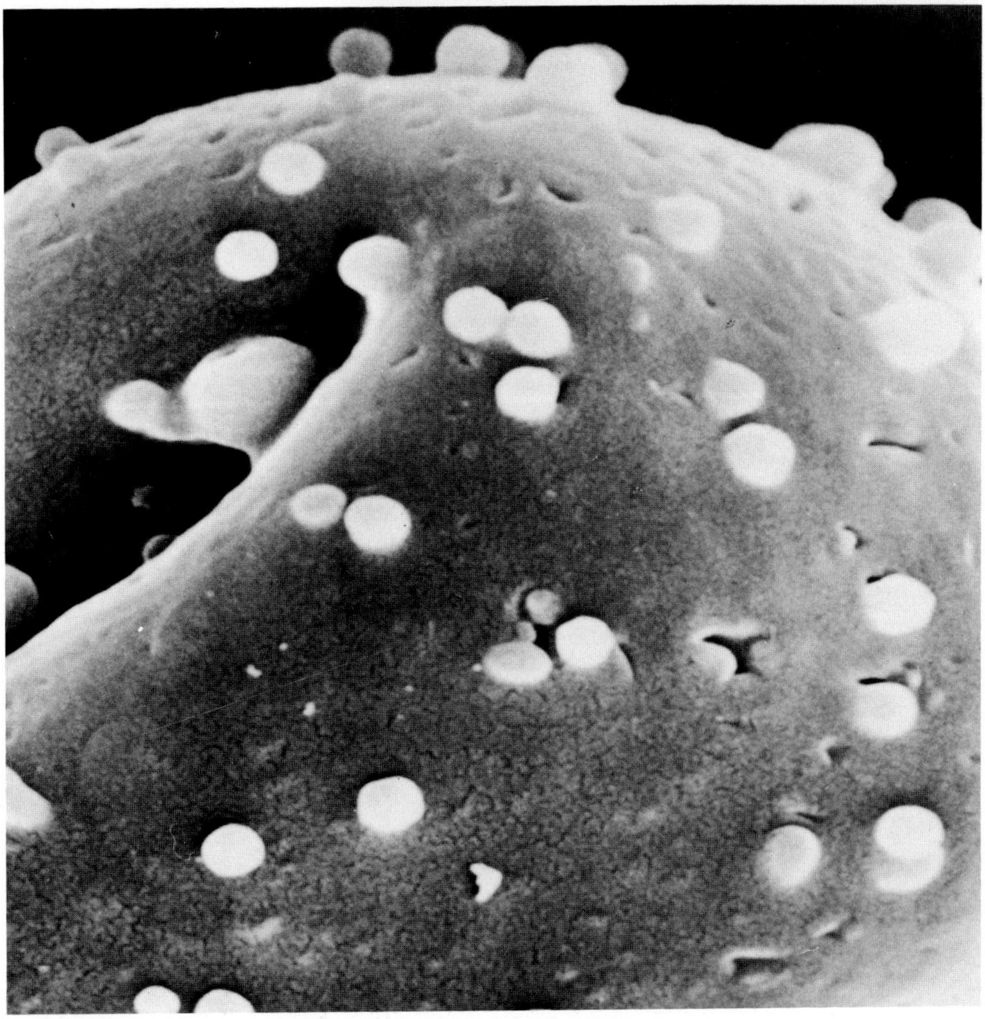

Fig. 2.70 Pollen morphology (48).
Top: *Fatsia japonica* Decnc. et Planch. [**Araliaceae**] (×1700).
Bottom: Detail of ornamentation (×8700).

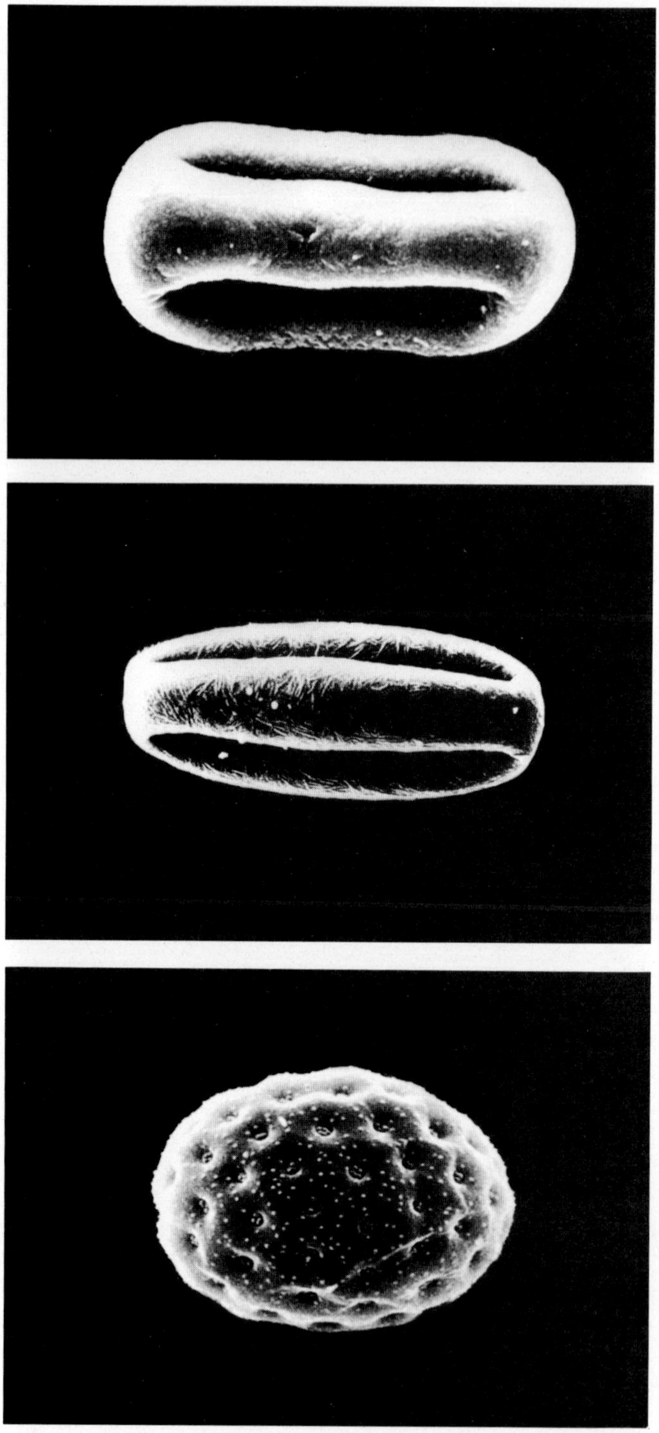

Fig. 2.71 Pollen morphology (49).
Top: Hedge-parsley (*Torilis japonica* DC.) [**Umbelliferae**] (×2200).
Middle: Sanicle (*Sanicula chinensis* Bunge) [**Umbelliferae**] (×1100).
Bottom: Red-Goose-foot (*Chenopodium album* L. var. *centrorubrum* Makino) [**Chenopodiaceae**] (×1700).

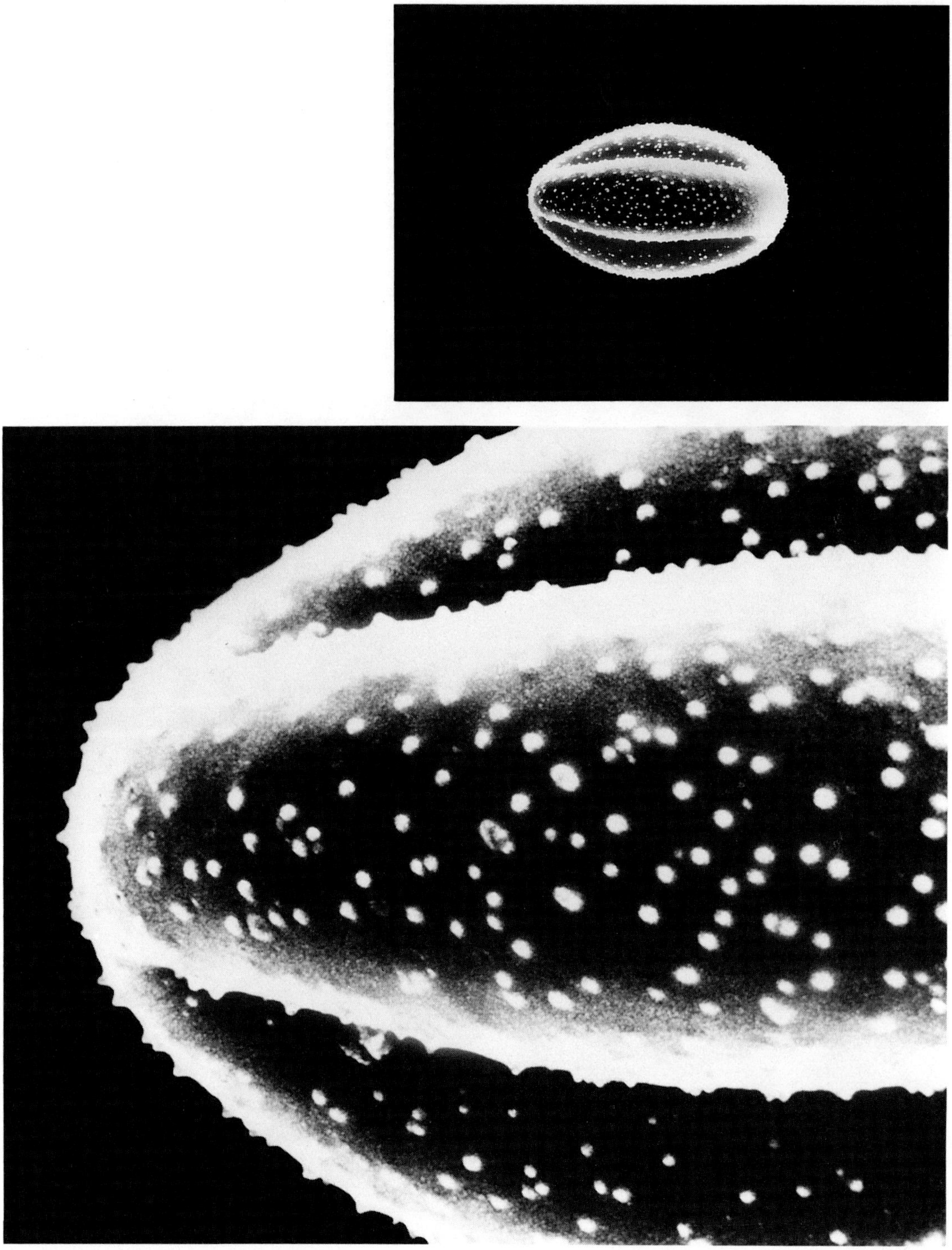

Fig. 2.72 Pollen morphology (50).
Top: Ink-bush (*Phytolacca americana* L.) [**Phytolaccaceae**] (×980).
Bottom: Detail of ornamentation (×52000).

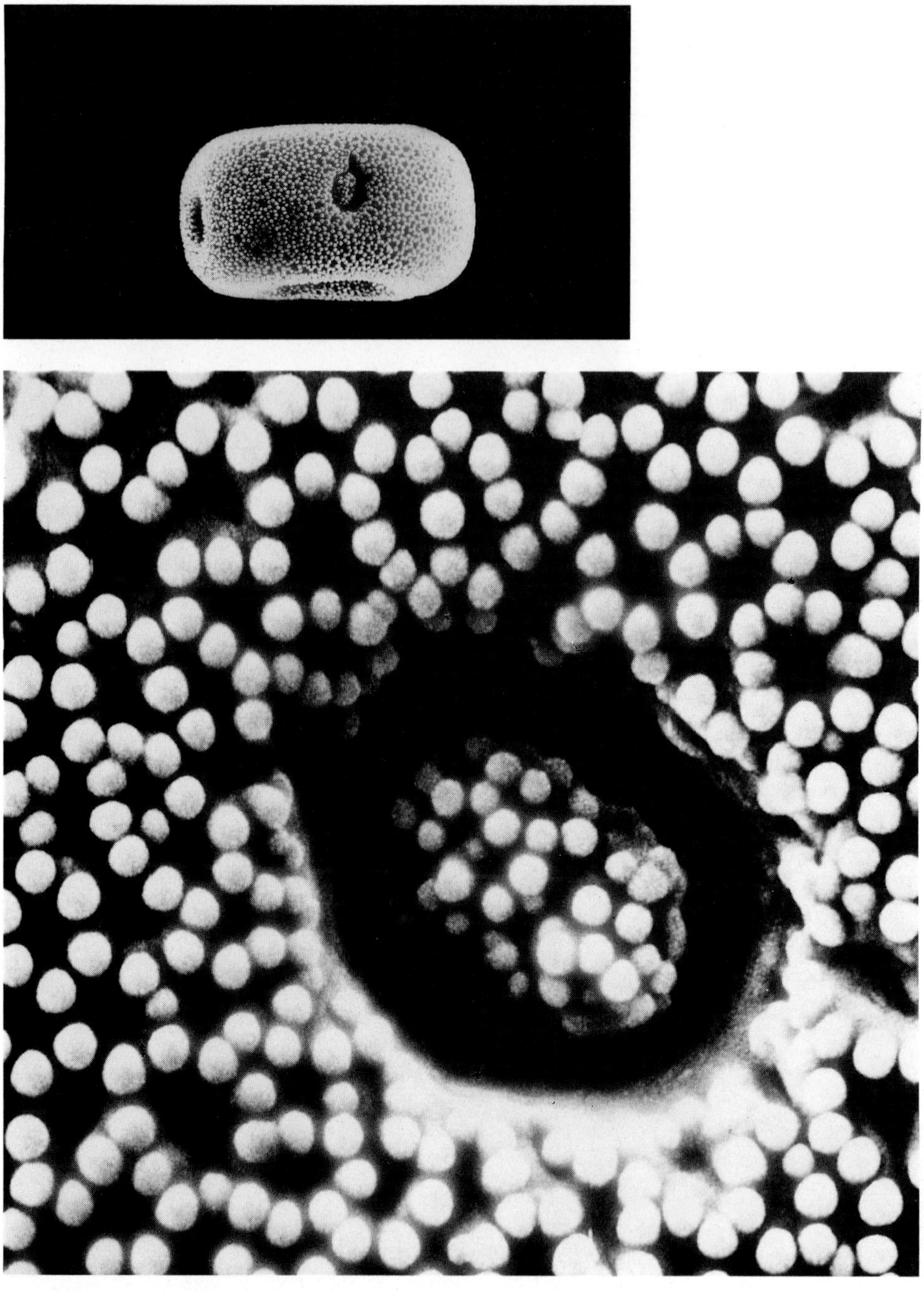

Fig. 2.73 Pollen morphology (51).
Top: Yamatogusa (*Cynocrambe japonica* Makino [**Theligonaceae**] (×1600)
Bottom: Detail of ornamentation (×24400).

Fig. 2.74 Pollen morphology (52).
Surface of pollen from Four-O'clock (*Mirabilis jalapa* L.) [**Nyctaginaceae**] (×5900).

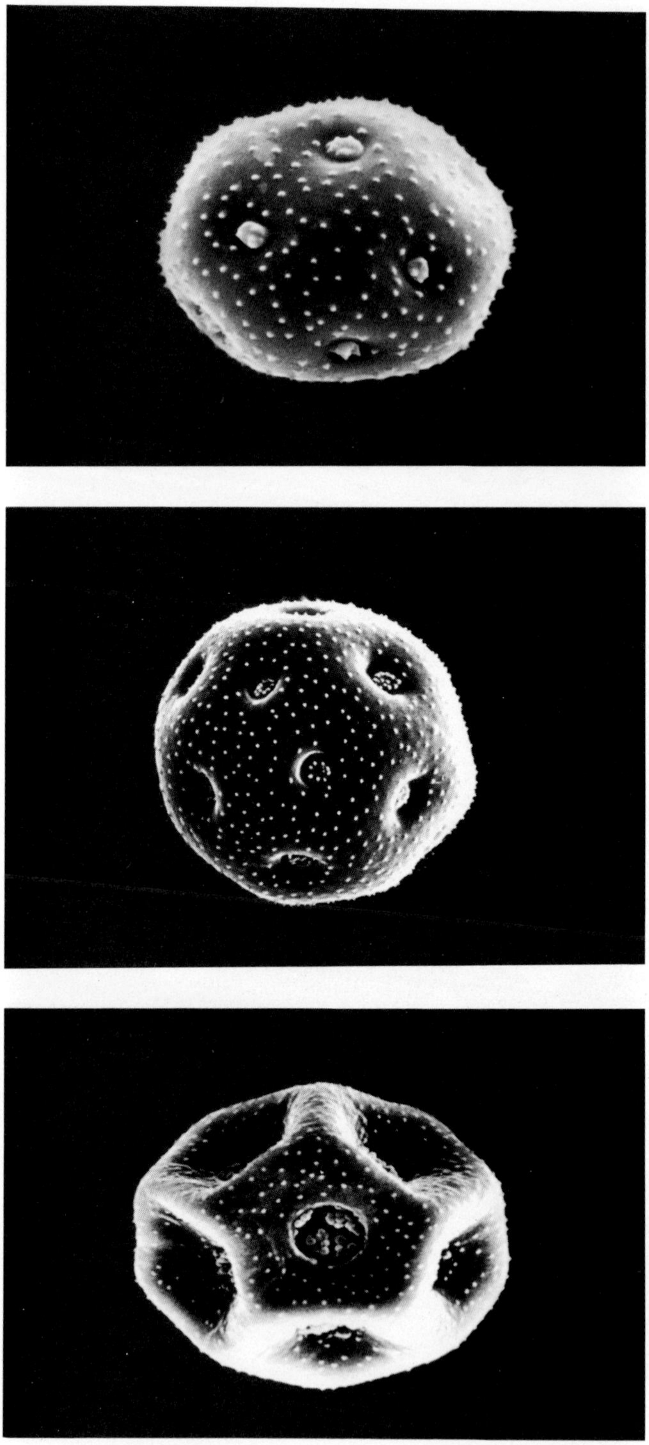

Fig. 2.75 Pollen morphology (53).
Top: Sweet-William (*Dianthus barbatus* L.) **[Caryophyllaceae]** (×1100).
Middle: Garden Pink (*Dianthus plumarius* L.) **[Caryophyllaceae]** (×800).
Bottom: Chickweed (*Stellaria media* L.) **[Caryophyllaceae]** (×1200).

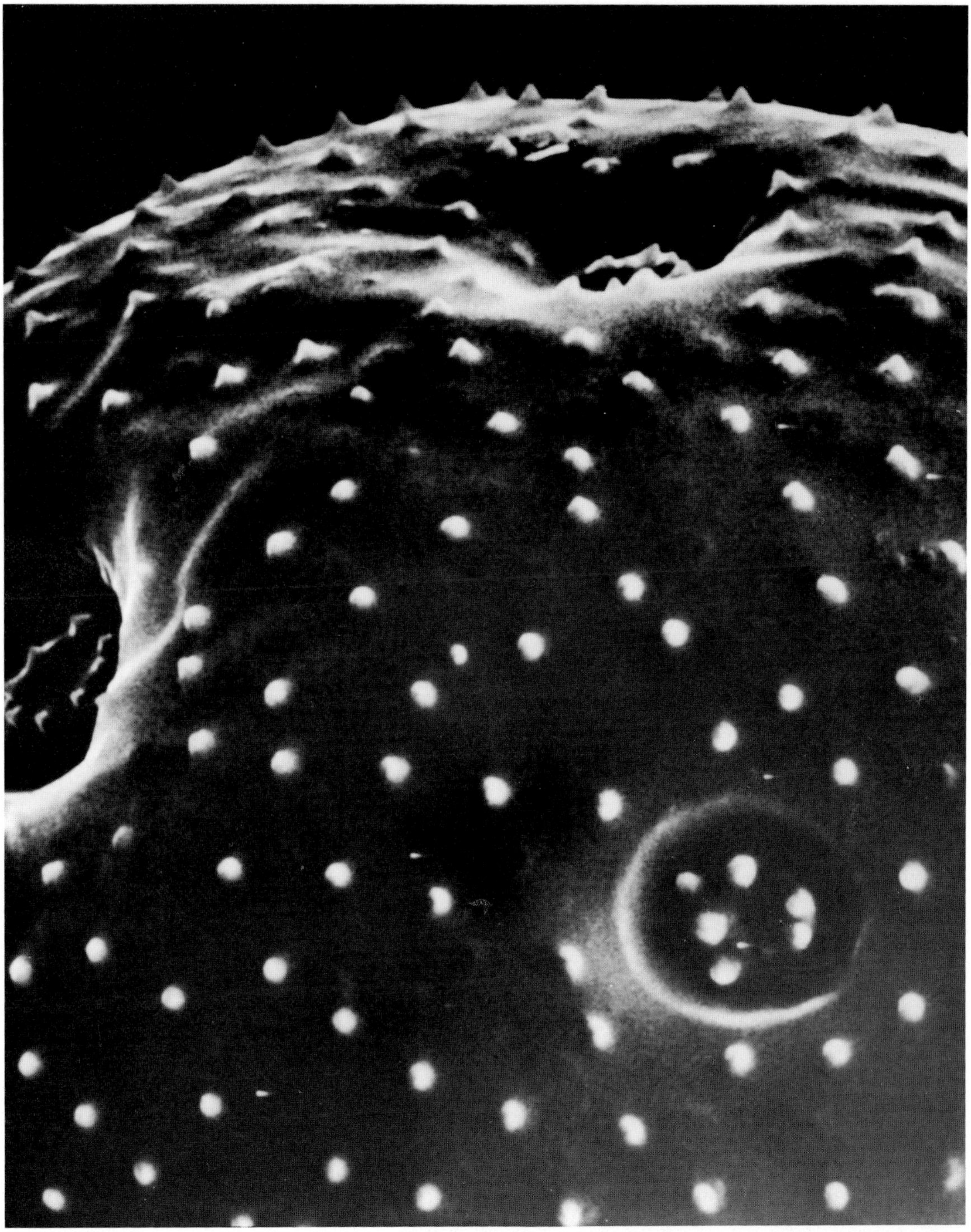

Fig. 2.76 Pollen morphology (54).
Surface of pollen from Cerast (*Cerastium Schmidtianum* Takeda) **[Caryophyllaceae]** (×7600).

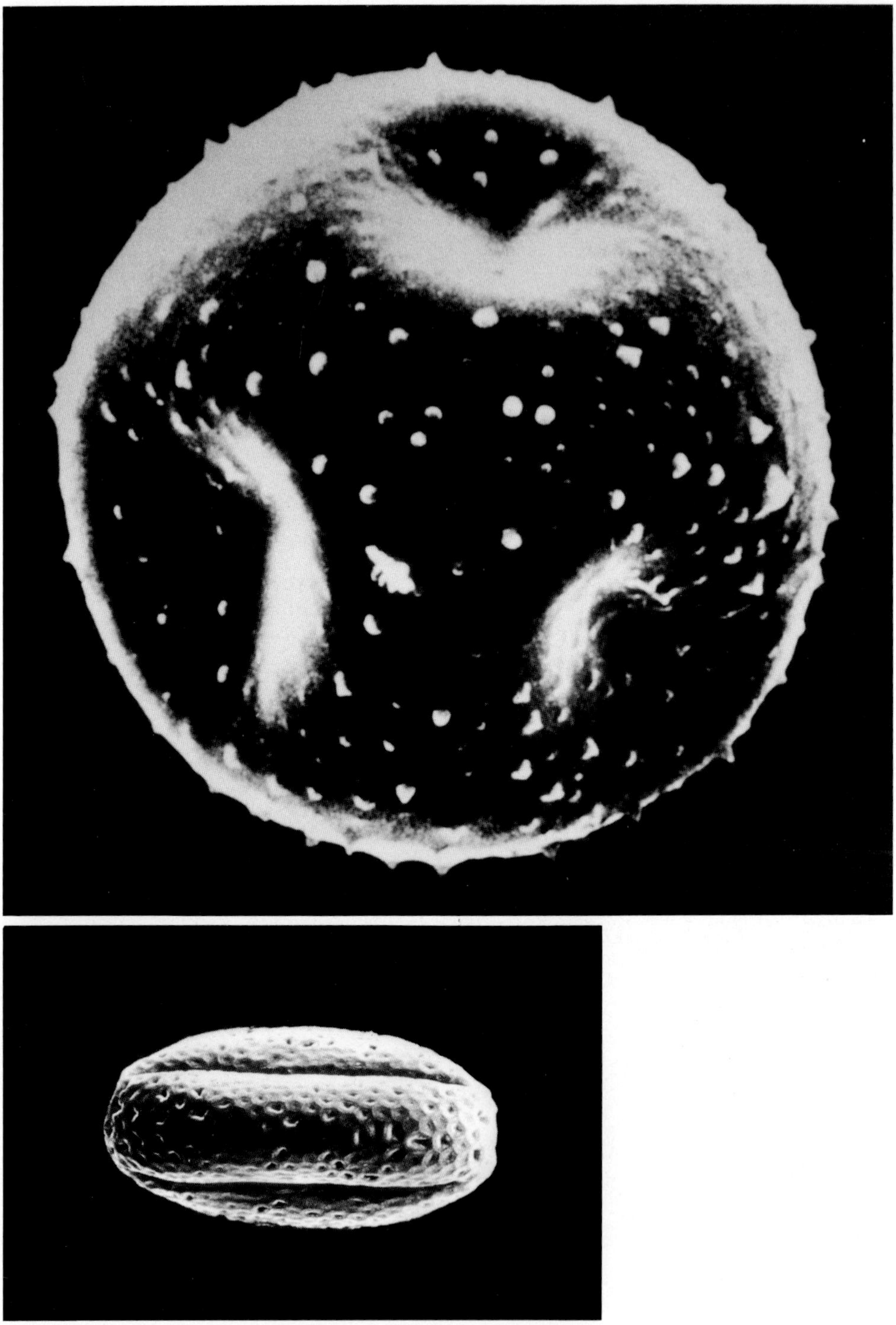

Fig. 2.77 Pollen morphology (55)
Top: Christmas cactus (*Zygocactus truncatus* Schum.) [**Cactaceae**] (×1000).
Bottom: Buckwheat (*Fagopyrum esculentum* Moench) [**Polygonaceae**] (×1000).

Fig. 2.78 Pollen morphology (56).
Prince Feather (*Polygonum orientale* L.) **[Polygonaceae]** ($\times$2000).

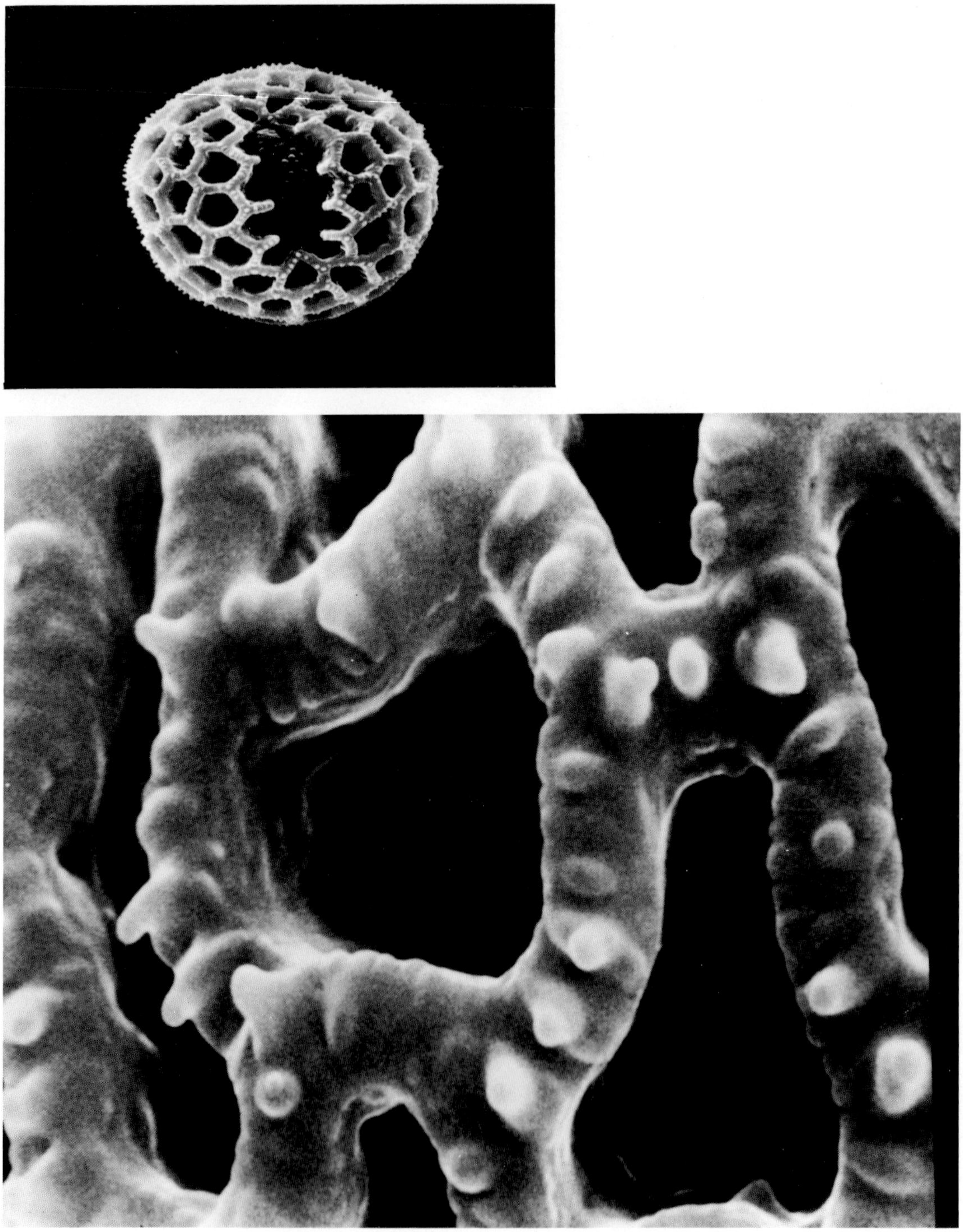

Fig. 2.79 Pollen morphology (57).
Top: Common Thrift (*Armeria vulgaris* Willd.) [**Plumbaginaceae**] (×700).
Bottom: Detail of ornamentation (×10000).

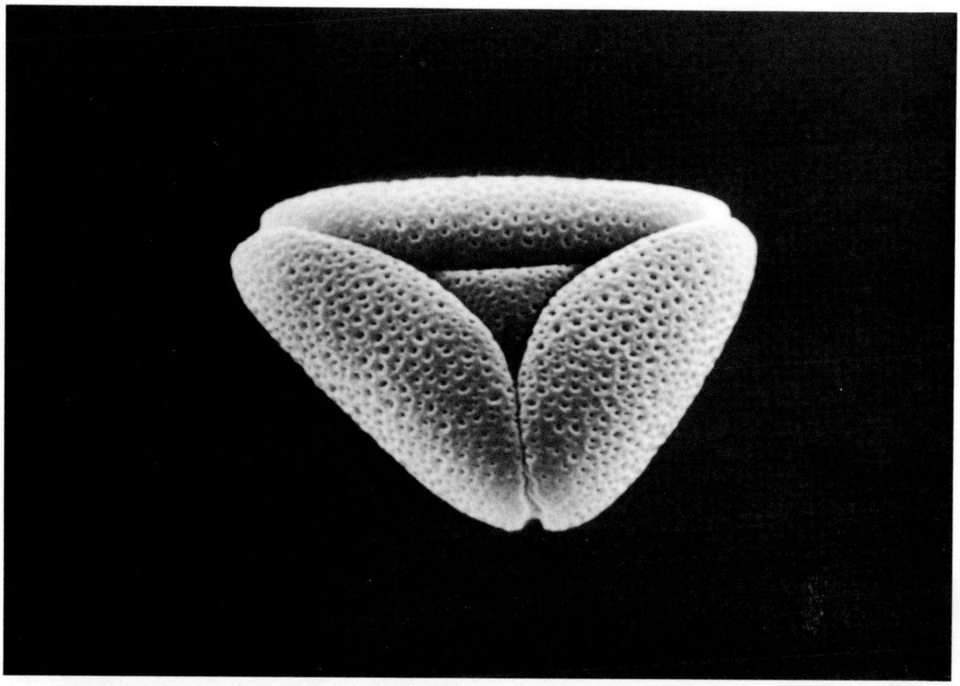

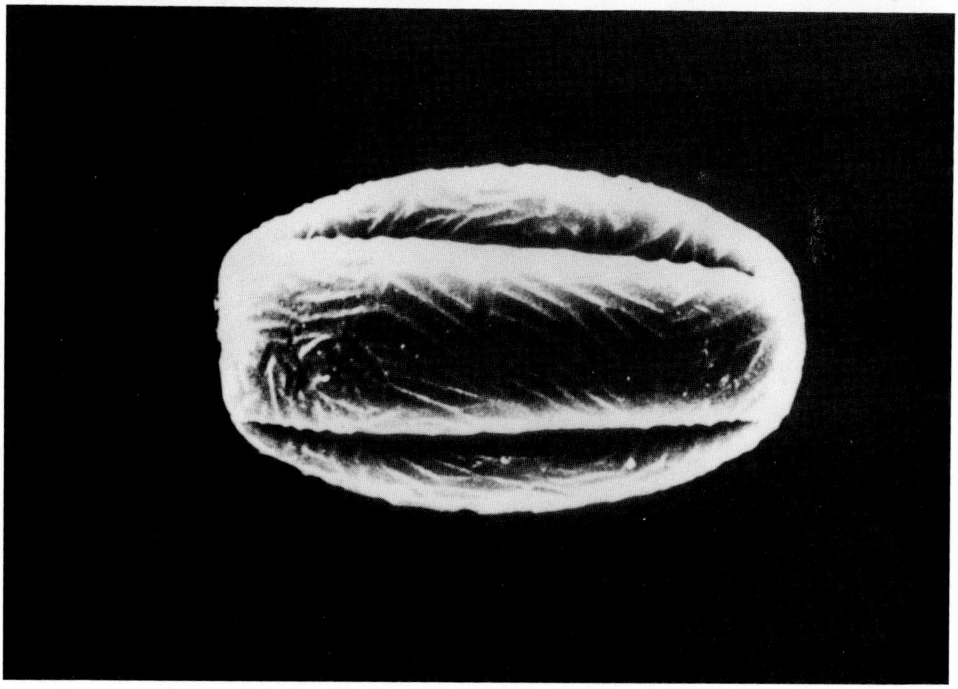

Fig. 2.80 Pollen morphology (58).
Top: Siebold's Primrose (*Primula Sieboldii* E. Morren) [**Primulaceae**] (×3600).
Bottom: *Buckleya lanceolata* Miq. [**Santalaceae**] (×2400).

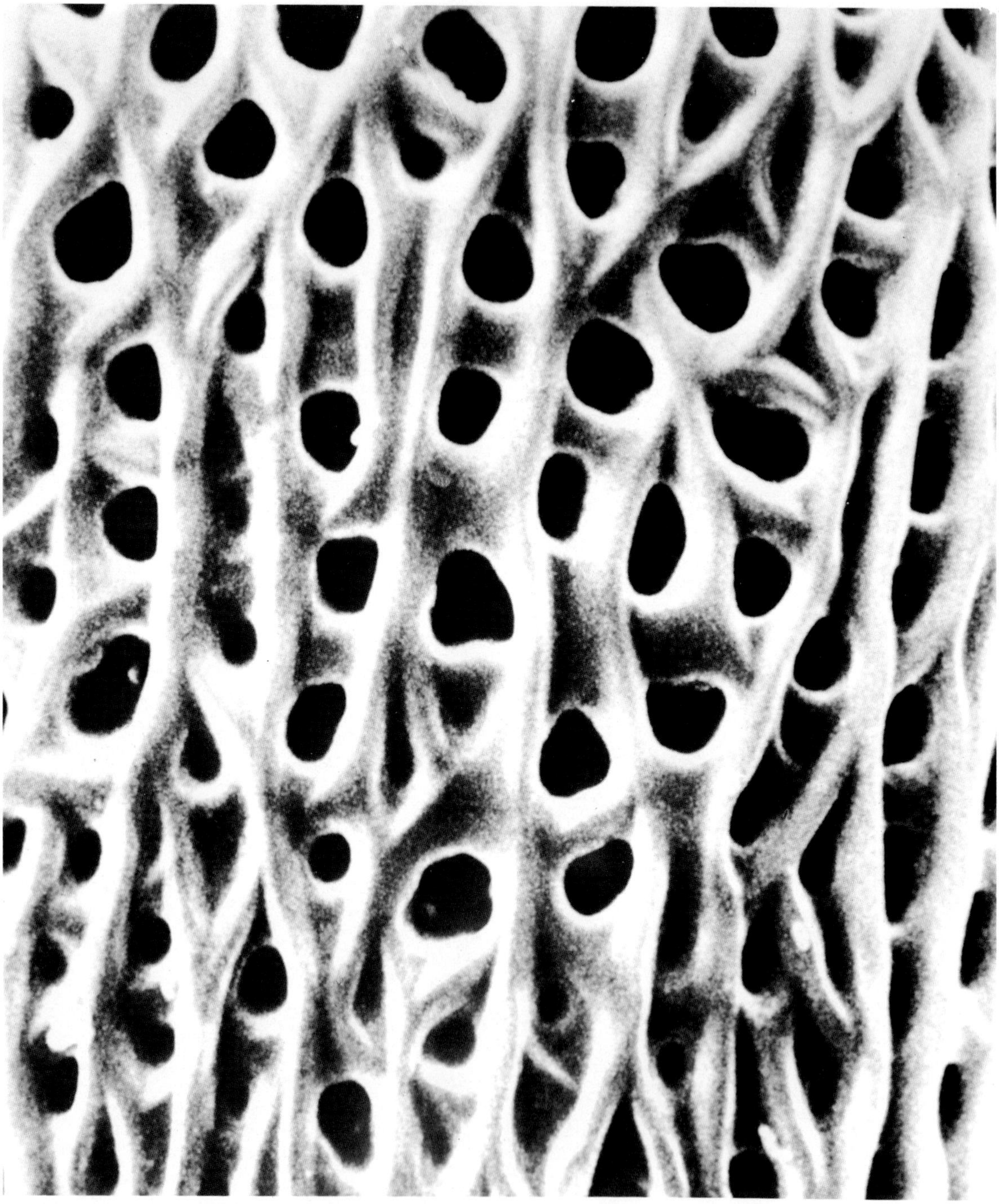

Fig. 2.81 Pollen morphology (59).
Surface of pollen from Gentian (*Gentiana Zollingeri* Fawc.) [**Gentianaceae**] (×20000).

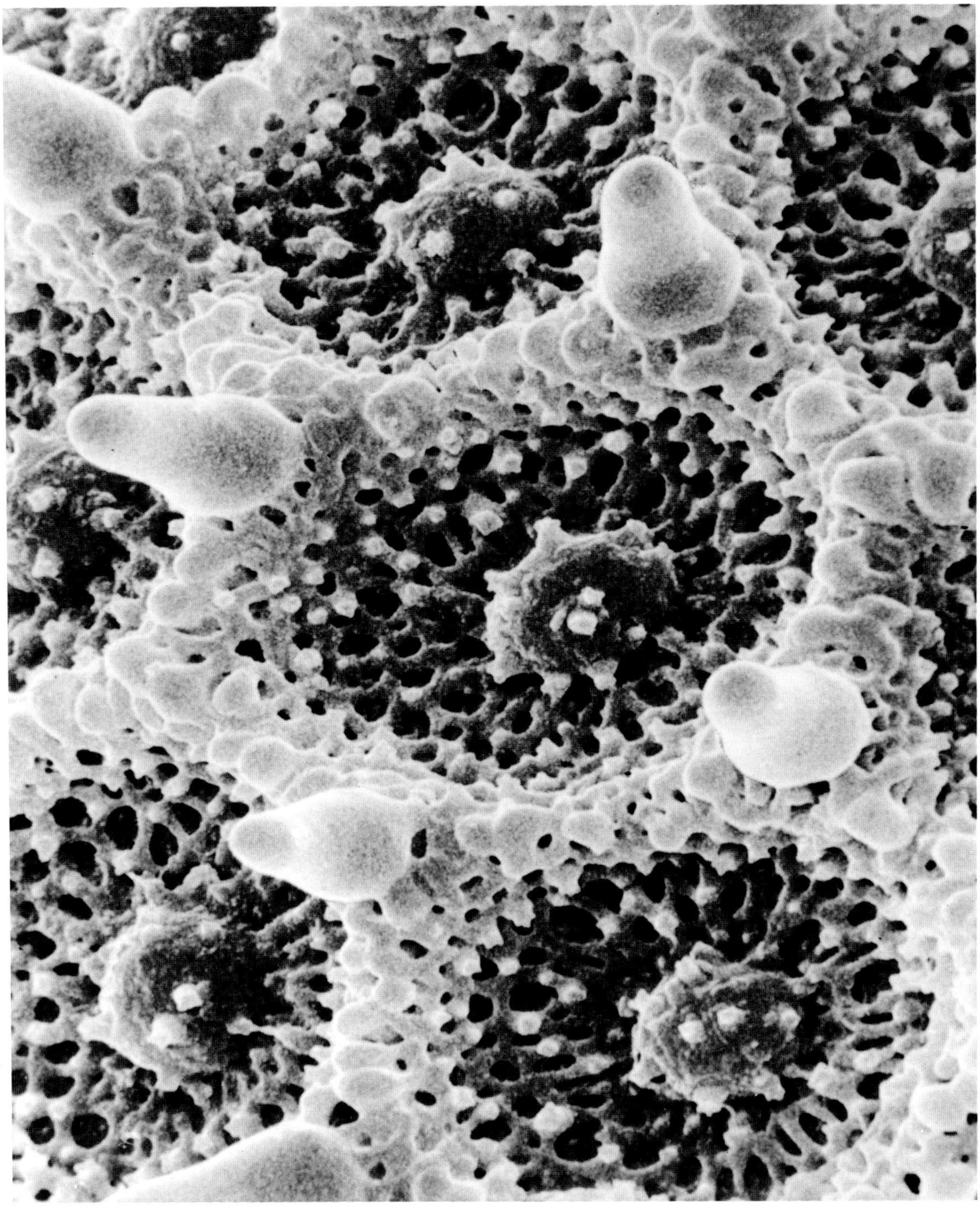

Fig. 2.82 Pollen morphology (60).
Surface of pollen from common Morning-glory (*Pharbitis nil Choisy*) **[Convolvulaceae]** ($\times$ 5000).

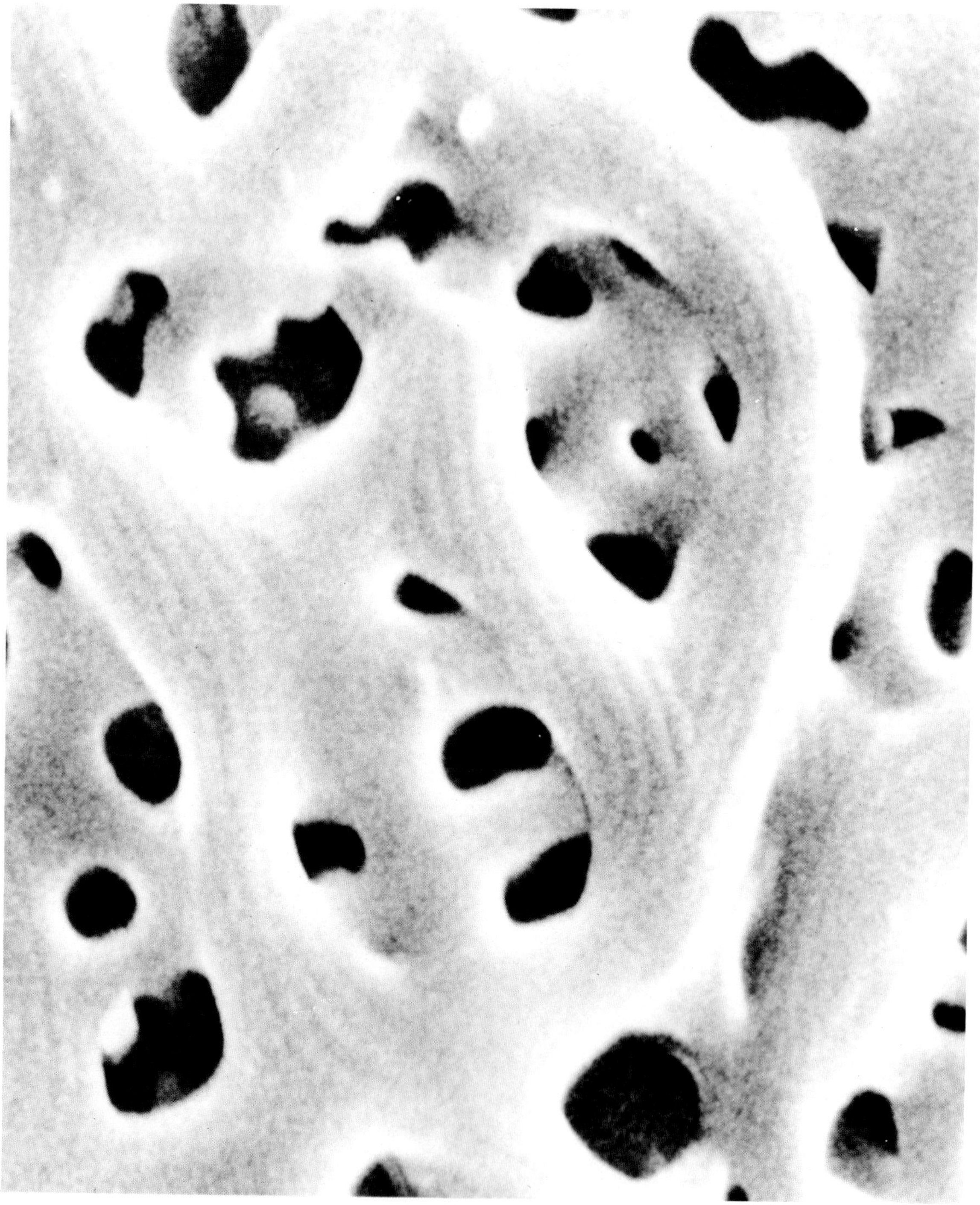

Fig. 2.83 Pollen morphology (61).
Surface of pollen from Stramonium (*Datura metel* L.) **[Solanaceae]** (×48000).

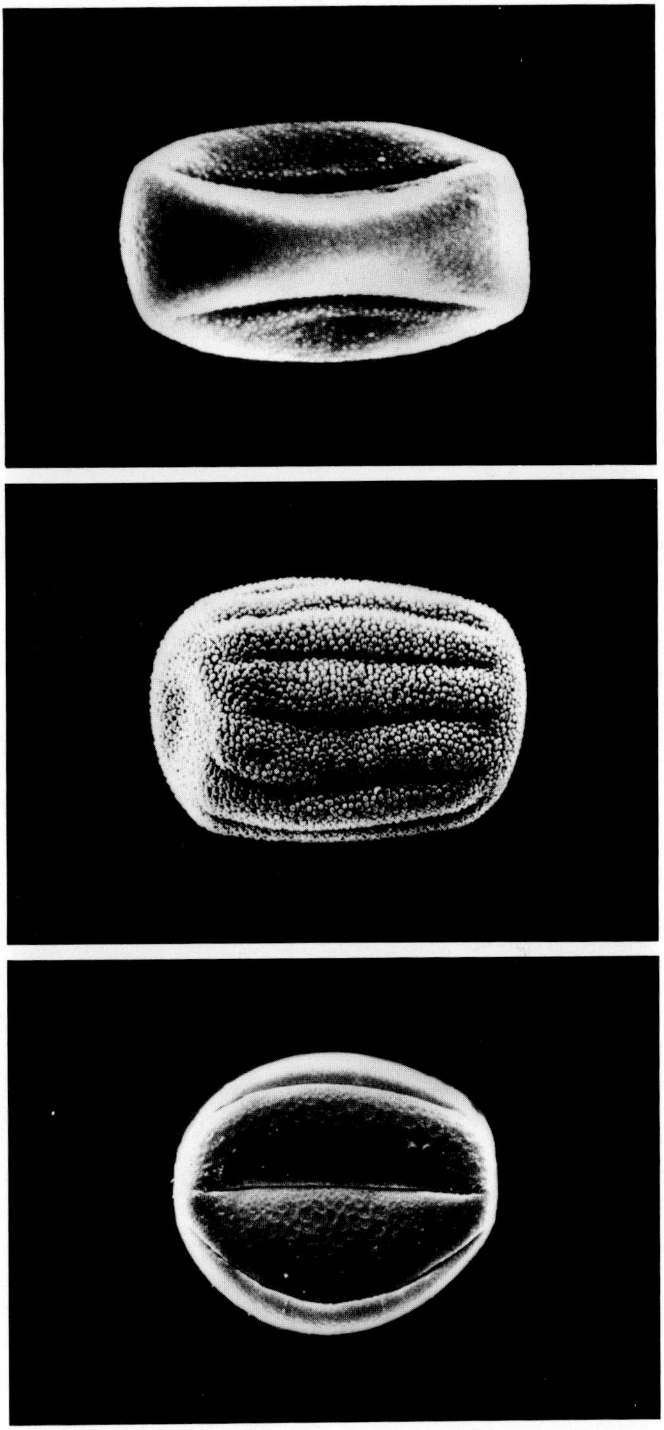

Fig. 2.84 Pollen morphology (62).
Top: *Conandron ramondioides* Sieb. et Zucc. [**Gesneriaceae**] (×4100).
Middle: Sesame (*Sesamum indicum* L.) [**Pedaliaceae**] (×800).
Bottom: *Rabdosia effusa* Hara [**Labiatae**] (×800).

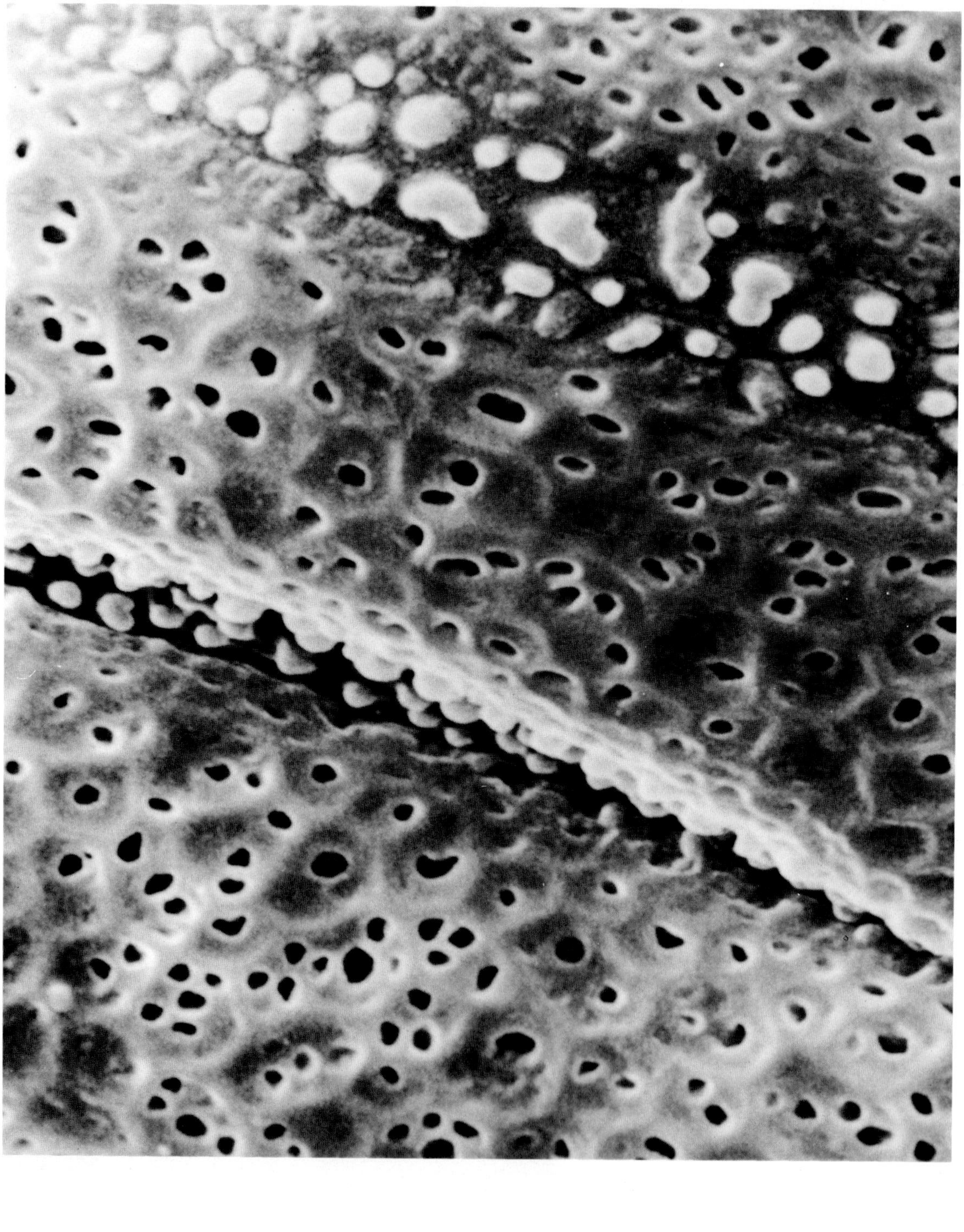

Fig. 2.85 Pollen morphology (63).
Surface of pollen from Sage (*Salvia officinalis* L.) [**Labiatae**] ($\times$5600).

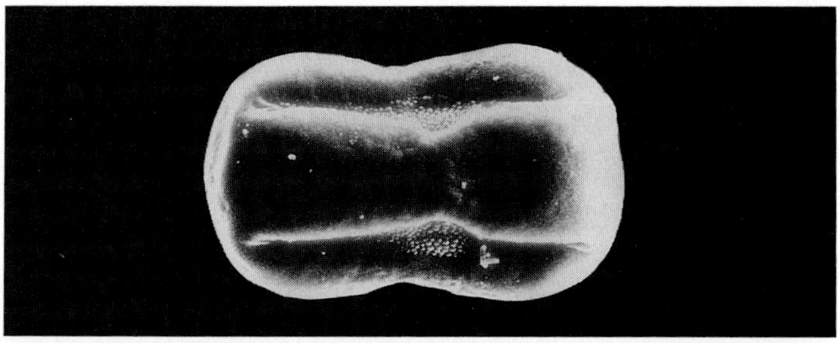

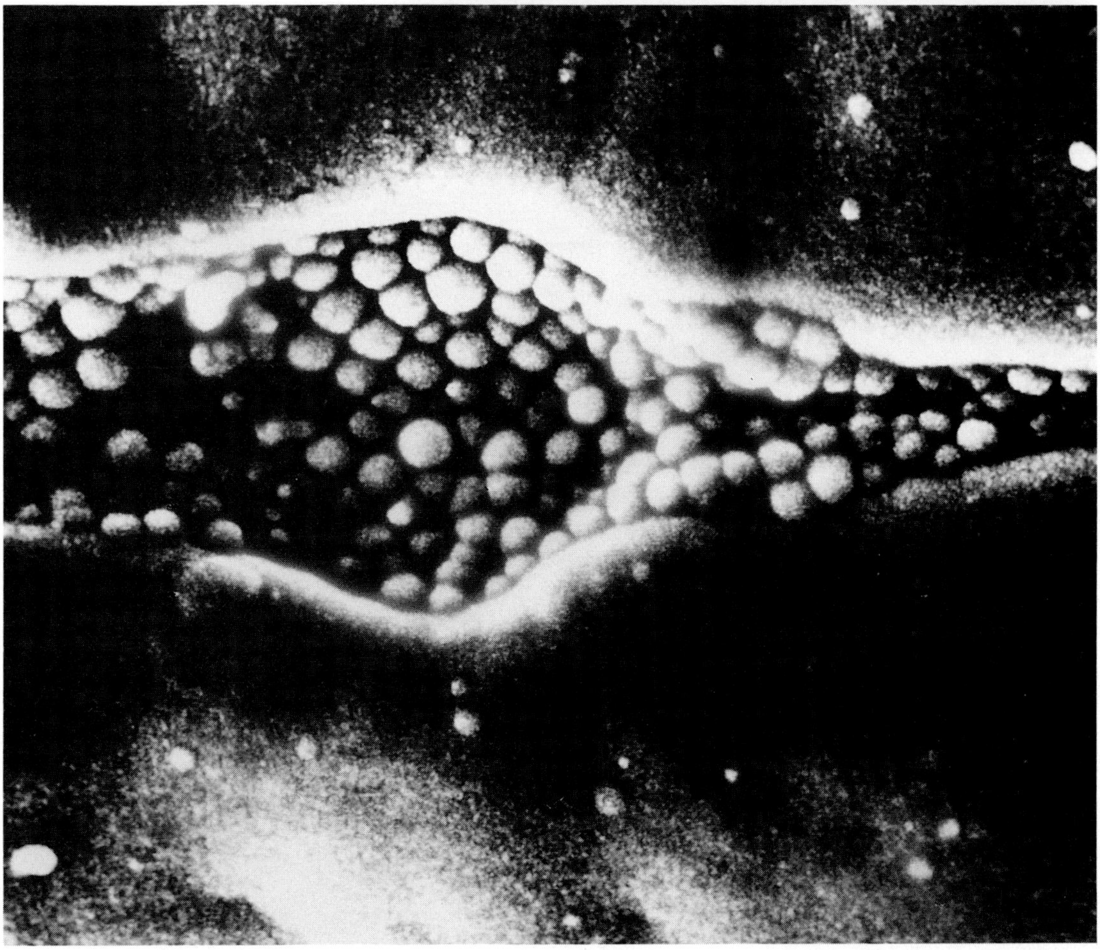

Fig. 2.86 Pollen morphology (64)
Top: Gromwell (*Lithospermum Zollingeri* A.DC.) **[Boraginaceae]** (×2600).
Bottom: Detail of ornamentation (×26000).

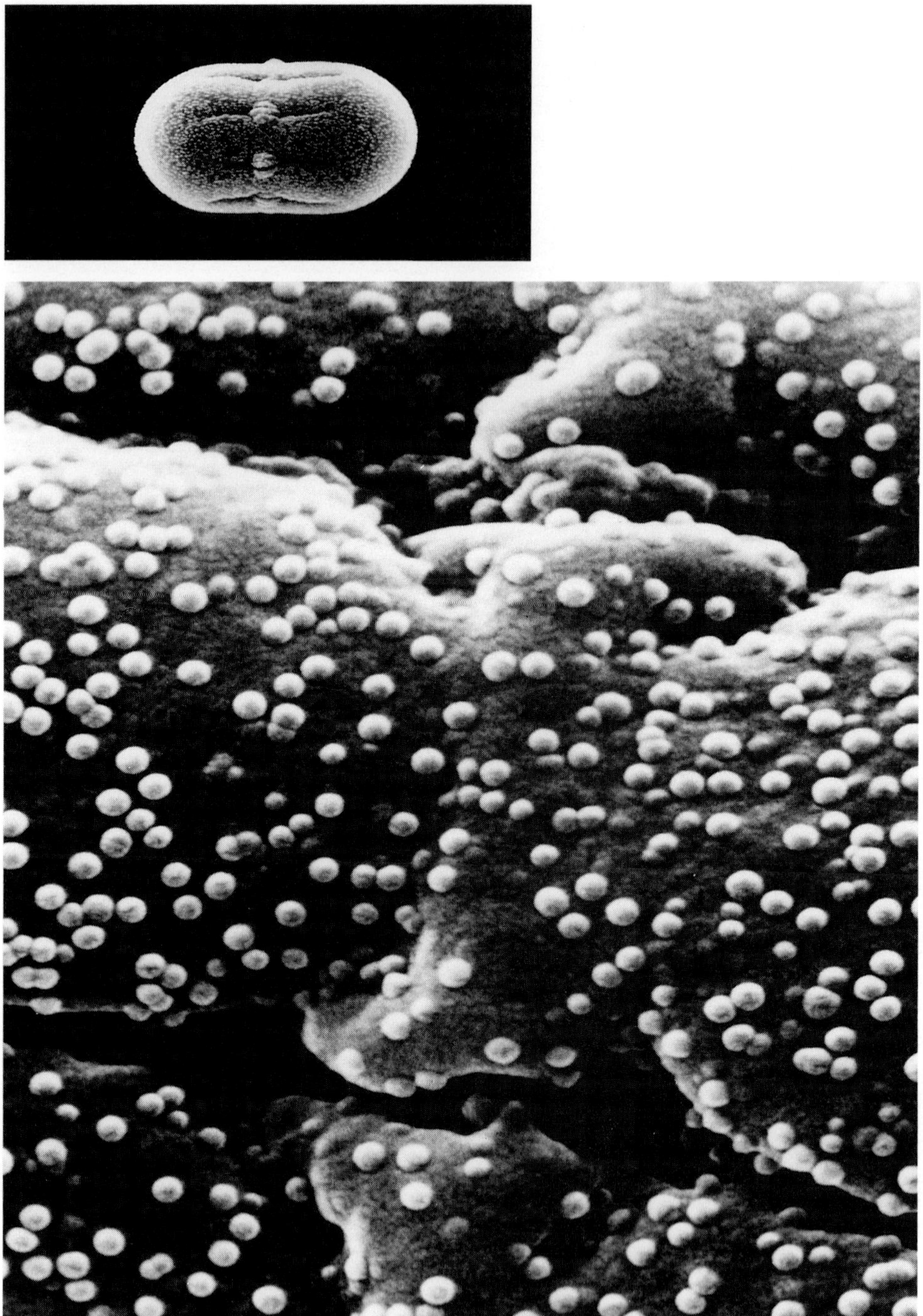

Fig. 2.87 Pollen morphology (65).
Top: Common comfrey (*Symphytum officinale* L.)[**Boraginaceae**](×1300).
Bottom: Detail of ornamentation (×16250).

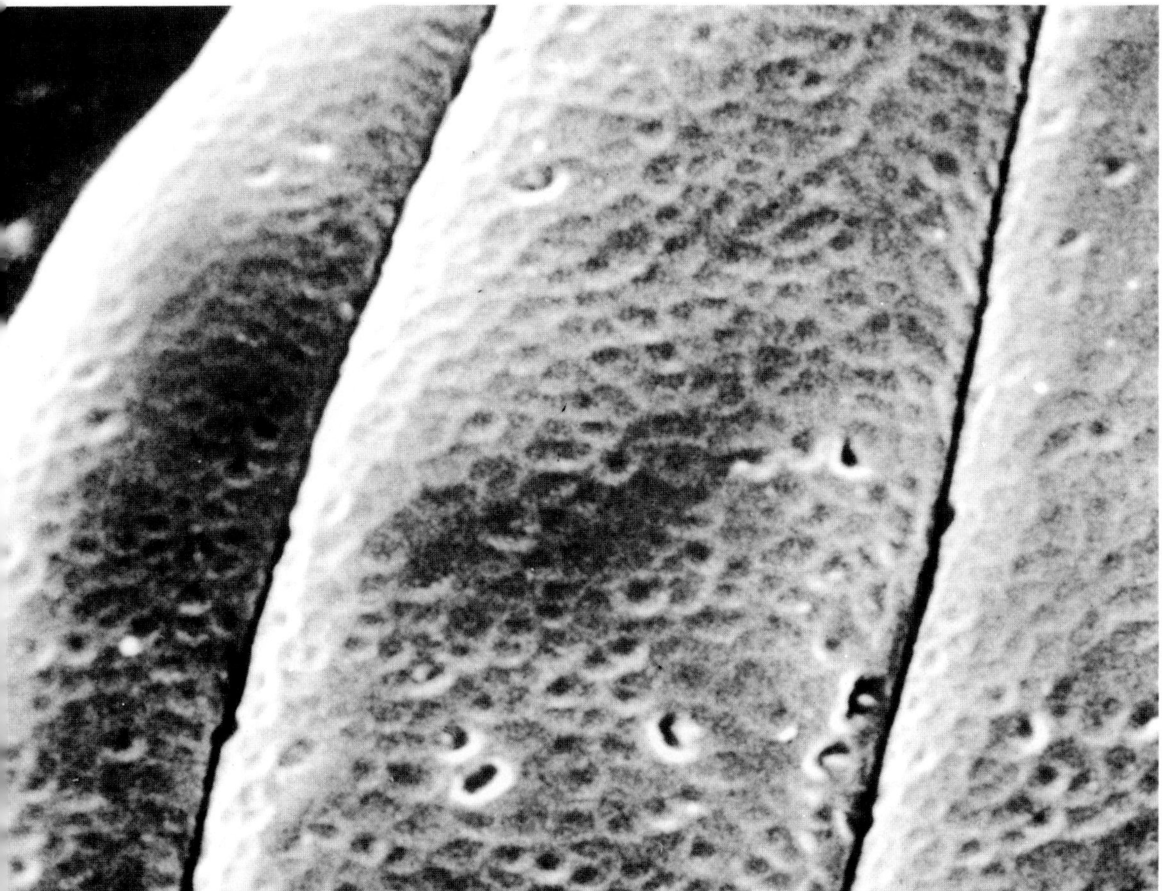

Fig. 2.88 Pollen morphology (66).
Surface of pollen from Ground Ivy (*Glechoma hederacea* L.) **[Labiatae]** (×7200).

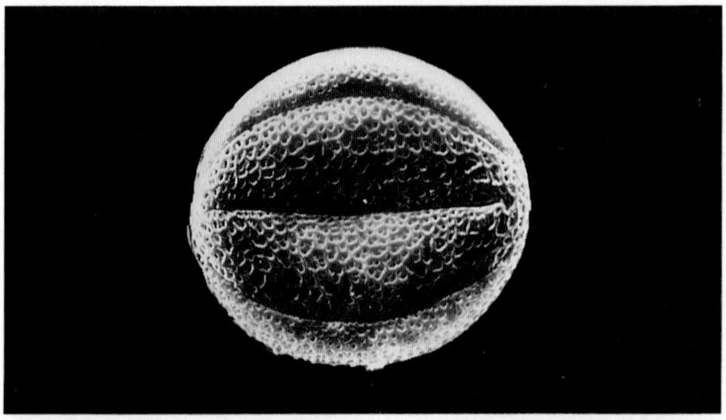

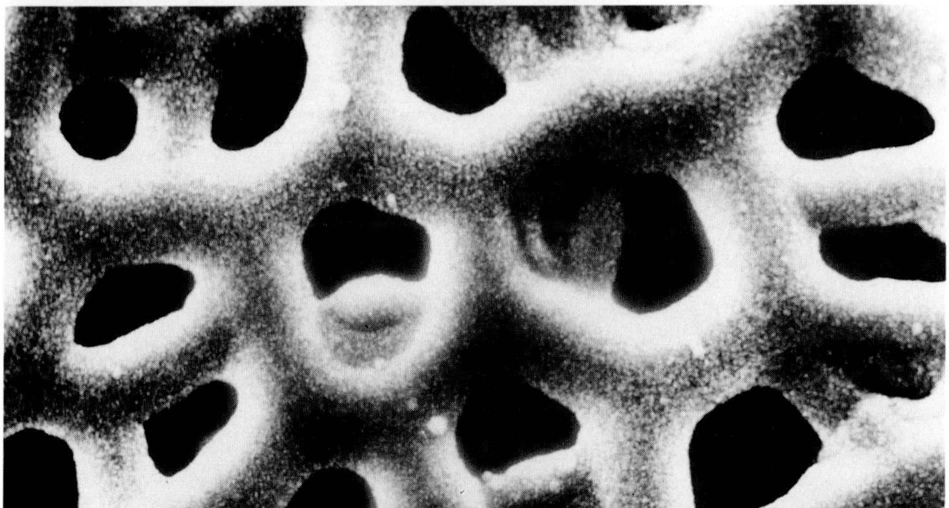

Fig. 2.89 Pollen morphology (67)
Top: Thyme (*Thymus quinquecostatus* Čelak.) **[Labiatae]** (×37500).
Bottom: Detail of ornamentation (×24400).

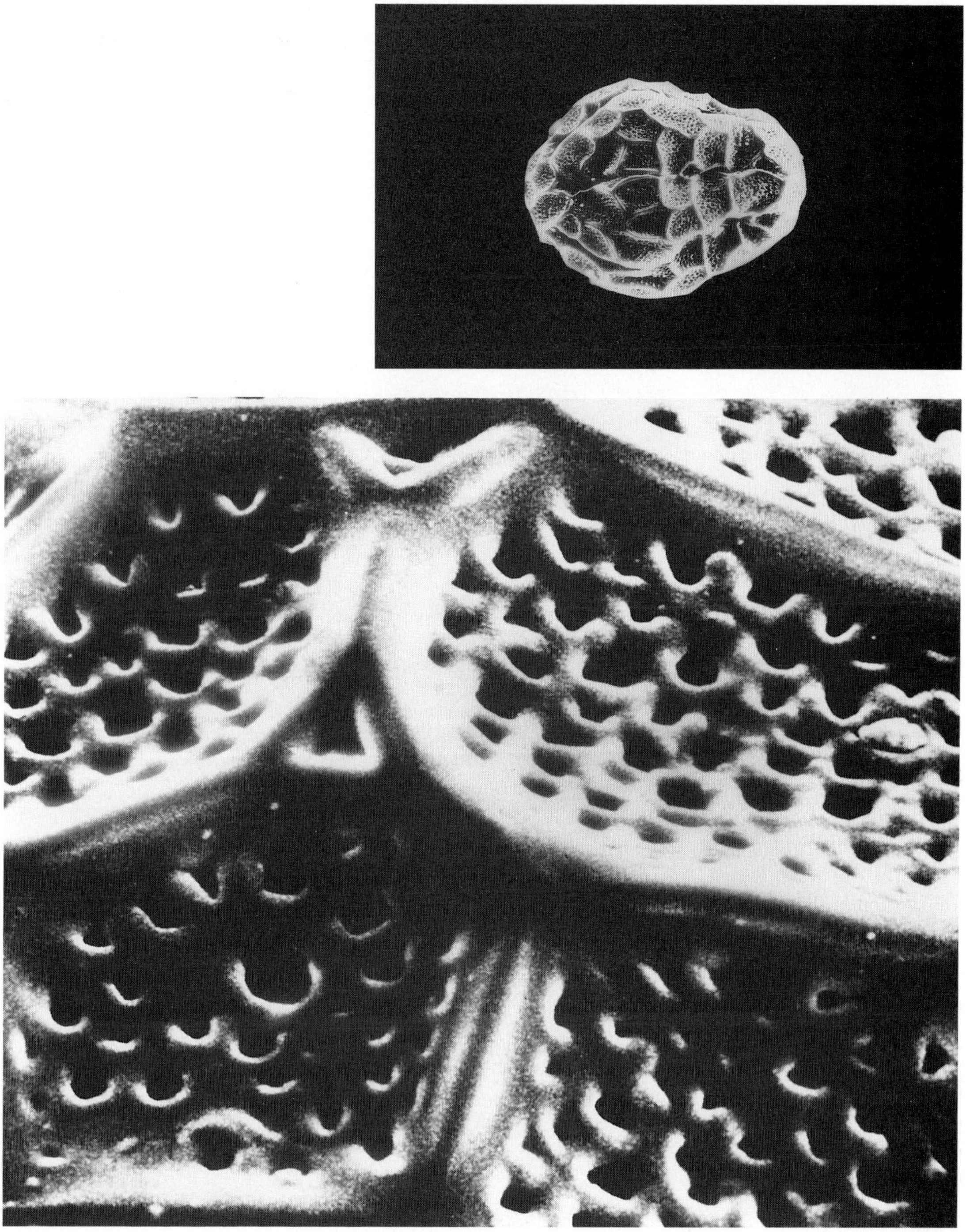

Fig. 2.90 Pollen morphology (68).
Top: Basil (*Ocimum basilicum*) [**Labiatae**] (×650).
Bottom: Detail of ornamentation (×12200).

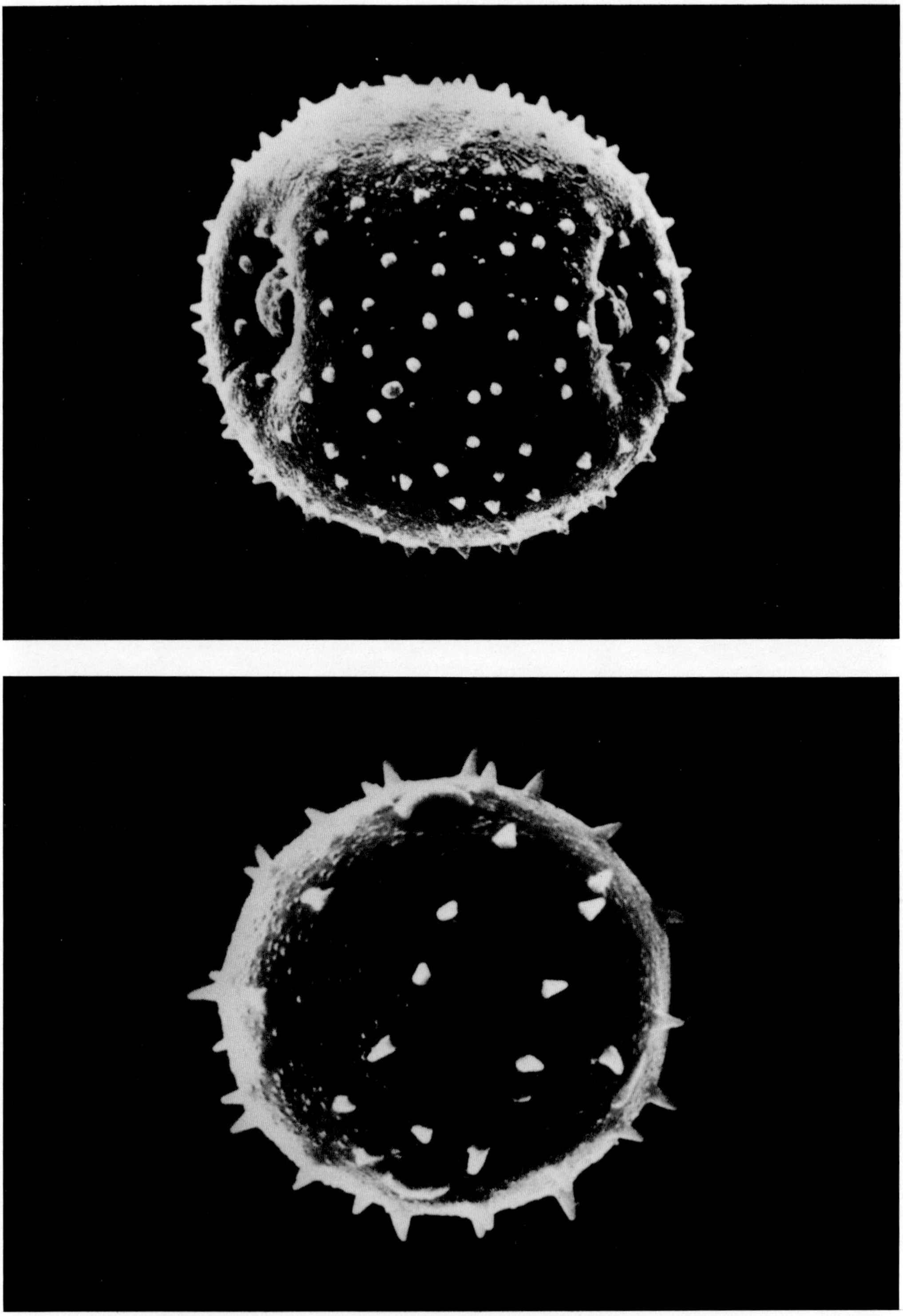

Fig. 2.91 Pollen morphology (69).
Top: Bellflower (*Campanula punctata* subsp. *hondoensis* Kitam.) **[Campanulaceae]** (×2000).
Bottom: Canterburry Bells (*Campanula medium* L.) **[Campanulaceae]** (×1500).

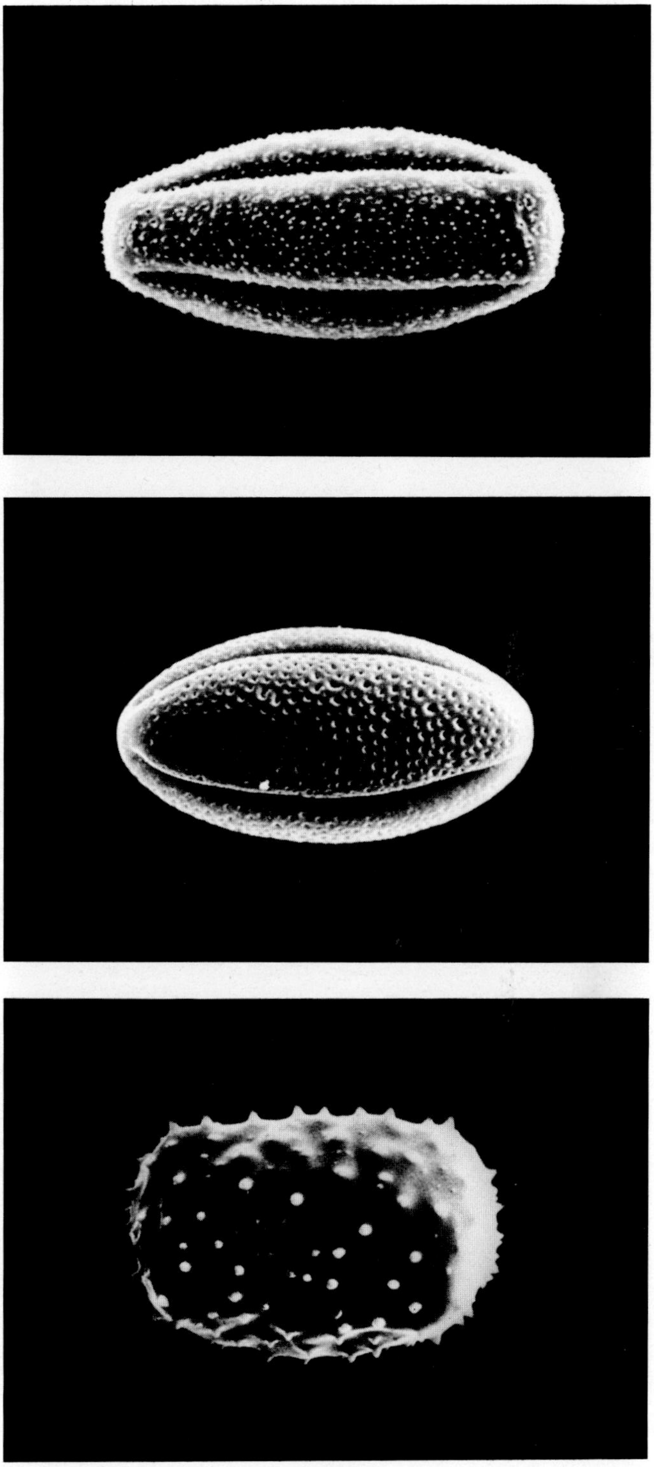

Fig. 2.92 Pollen morphology (70).
Top: *Paederia scandens* Merrill [**Rubiaceae**] (×1200).
Middle: Elder (*Sambucus Sieboldiana* Blume) [**Caprifoliaceae**] (×1700).
Bottom: *Patrinia villosa* Juss. [**Valerianaceae**] (×800).

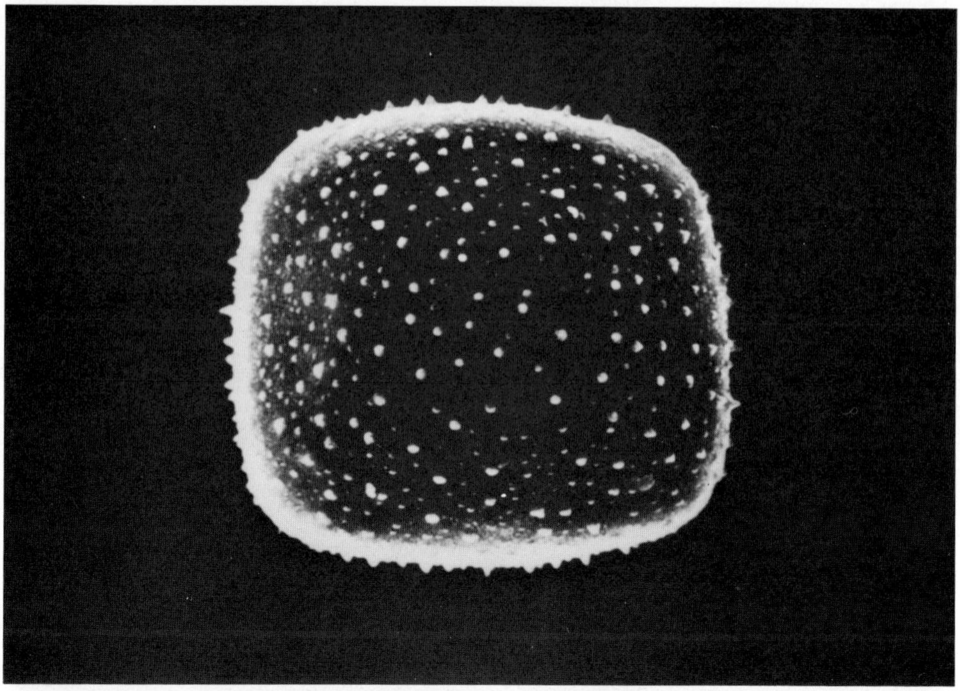

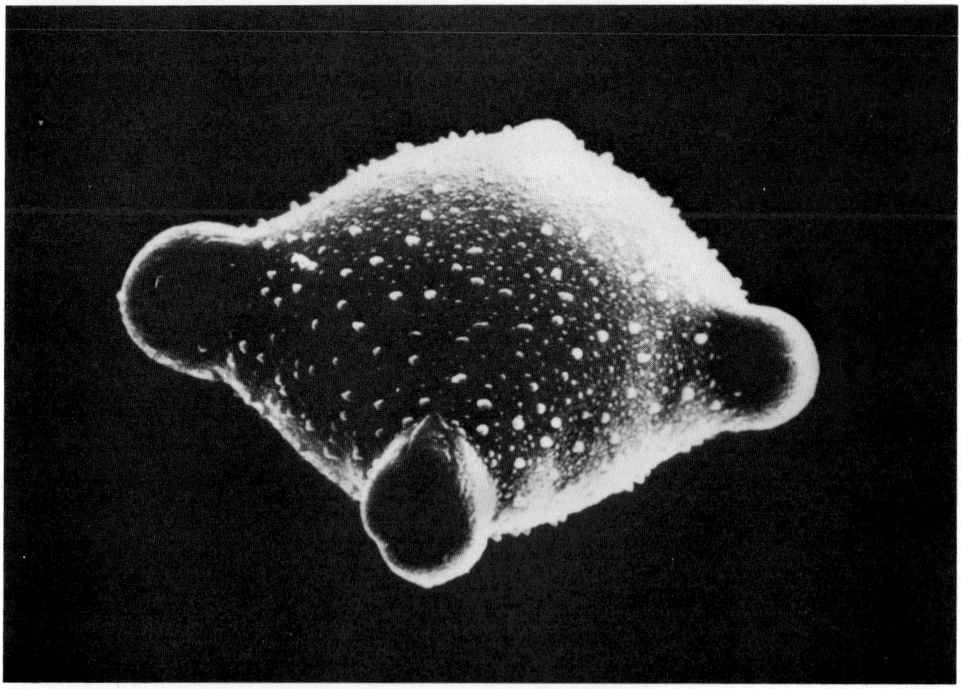

Fig. 2.93 Pollen morphology (71).
Top: Polar view of Honeysuckle (*Lonicera gracilipes* Miq. var. *glandulosa* Maxim.) **[Caprifoliaceae]** (×1200).
Bottom: Equatorial view of Honeysuckle (×1200).

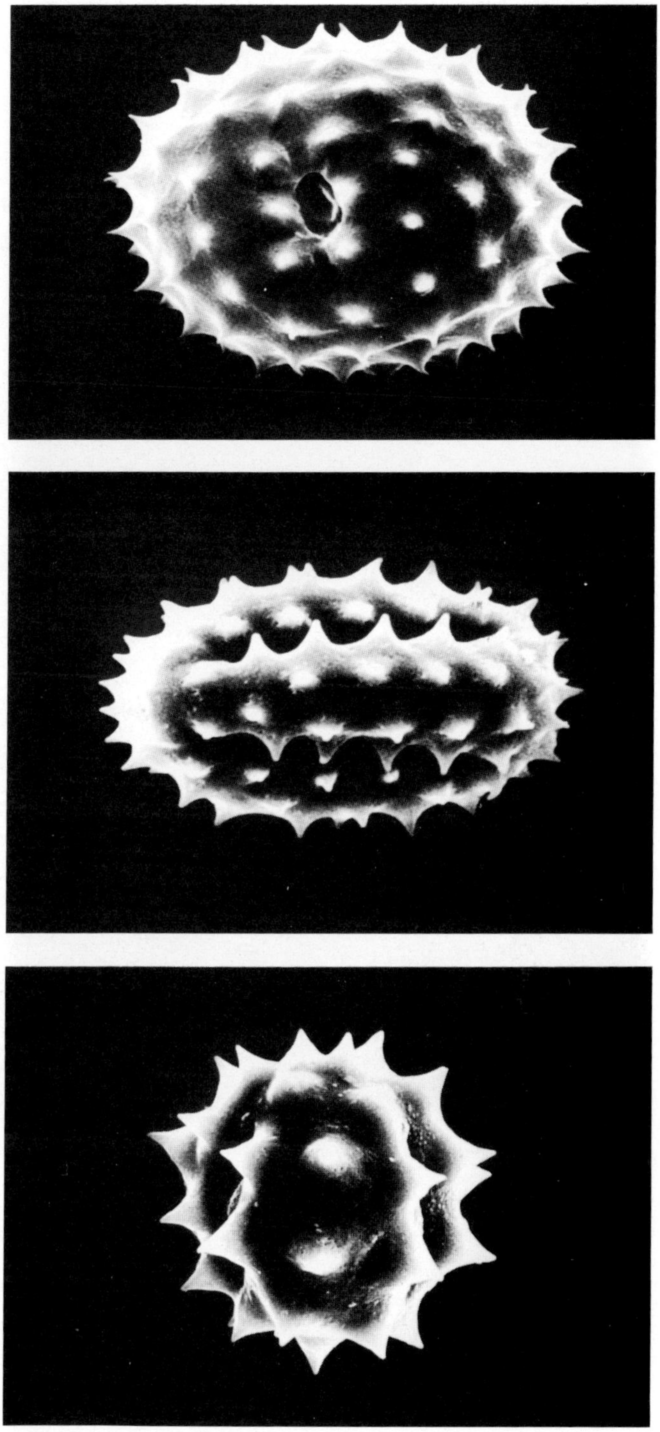

Fig. 2.94 Pollen morphology (72).
Top: Great-Ragweed (*Ambrosia trifida* L.) **[Compositae]** (×2200).
Middle: Thoroughwort (*Eupatorium japonicum* Thunb.) **[Compositae]** (×1700).
Bottom: Margaerite (*Chrysanthemum frutescens* L.) **[Compositae]** (×1700).

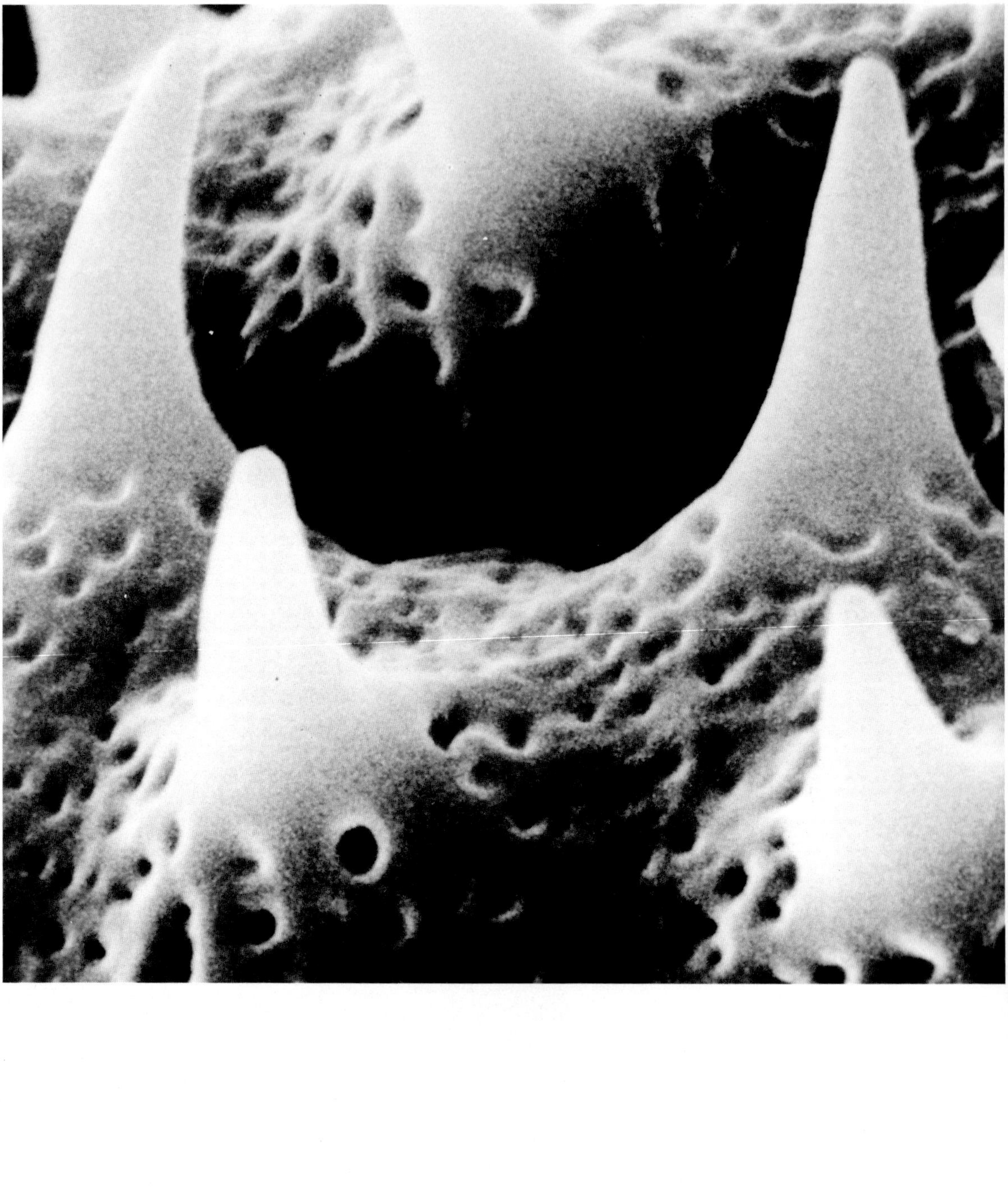

Fig. 2.95 Pollen morphology (73).
Surface of pollen from Muster John Henry (*Tagetes patula* L.) **[Compositae]** (×19000).

Fig. 2.96 Pollen morphology (74).
Top: *Ixeris debilis* A. Gray **[Compositae]** (×2400).
Bottom: Dandelion (*Taraxacum officinale* Weber) **[Compositae]** (×3200).

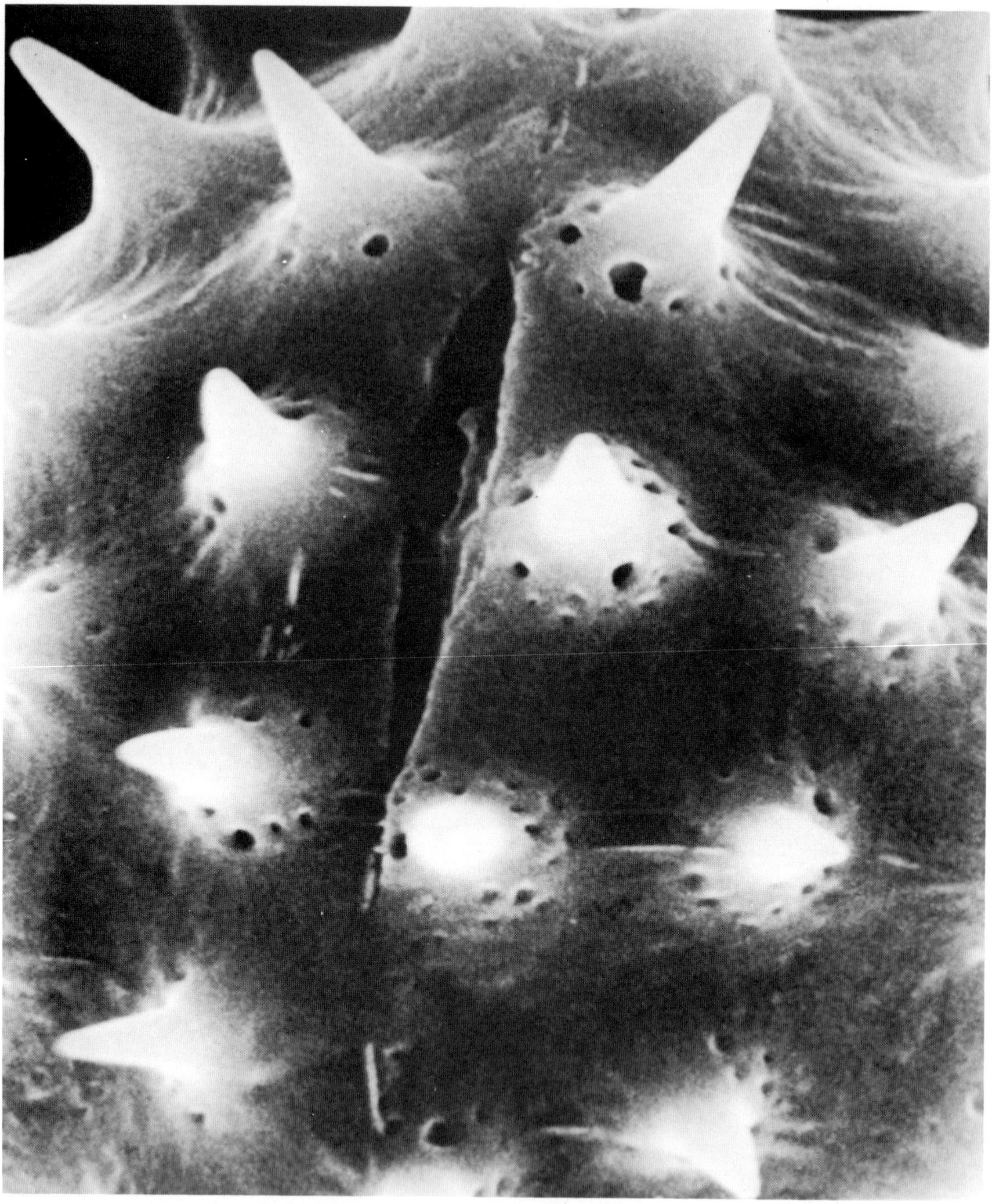

Fig. 2.97 Pollen morphology (75).
Surface of pollen from Cosmos (*Cosmos bipinnatus* Cav.) **[Compositae]** (×9400).

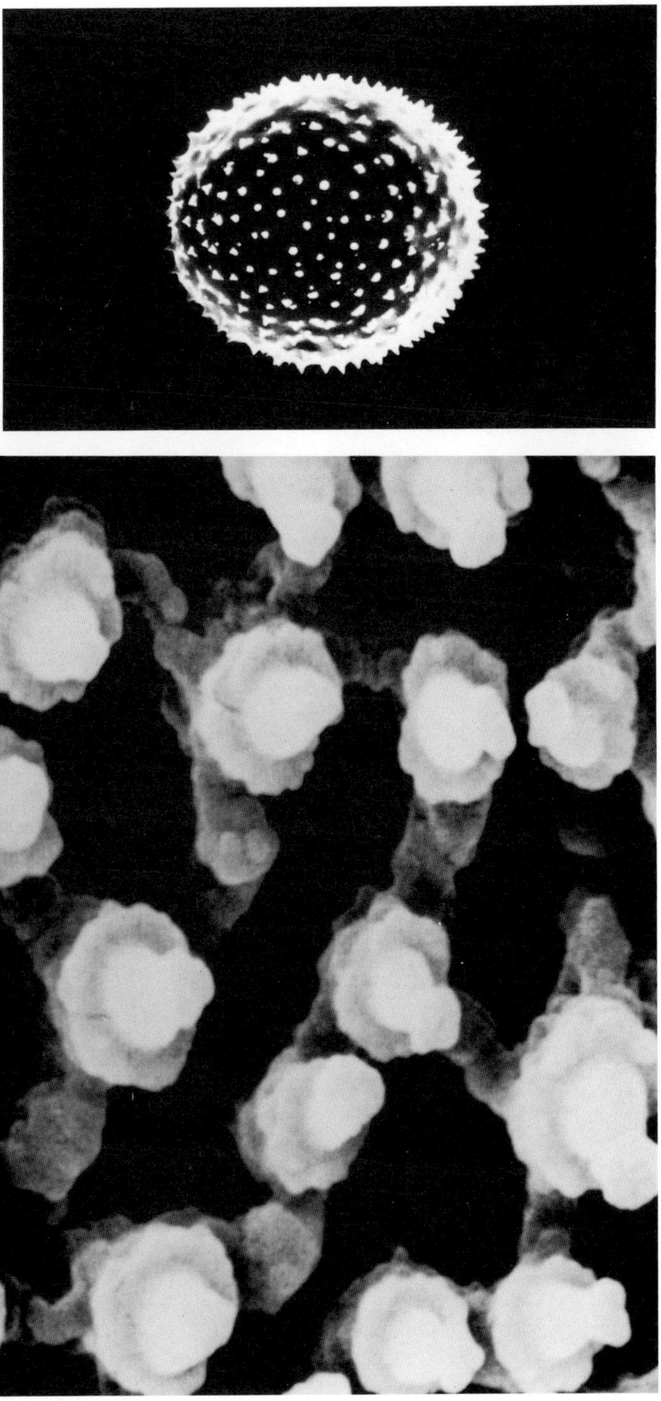

Fig. 2.98 Pollen morphology (76).
Top: Brazilian Elodea (*Egeria densa* St. Gohn) [**Hydrocharitaceae**] (×800).
Bottom: Surface of pollen from *Hydrilla verticillata* Casp. [**Hydrocharitaceae**] (×16000).

Fig. 2.99 Pollen morphology (77).
Surface of pollen from Ginger (*Zingiber Mioga* Roscoe) [**Zingiberaceae**] (×5000).

Fig. 2.100 Pollen morphology (78).
Top: Maximowicz's Lily (*Lilium Maximowiczii* Regel) **[Liliaceae]** (×800).
Midde: *Disporum smilacinum* A. Gray **[Liliaceae]** (×800).
Bottom: Lily of the Valley (*Convallaria majalis* L.C. Keiskei Miquel) **[Liliaceae]** (×1300).

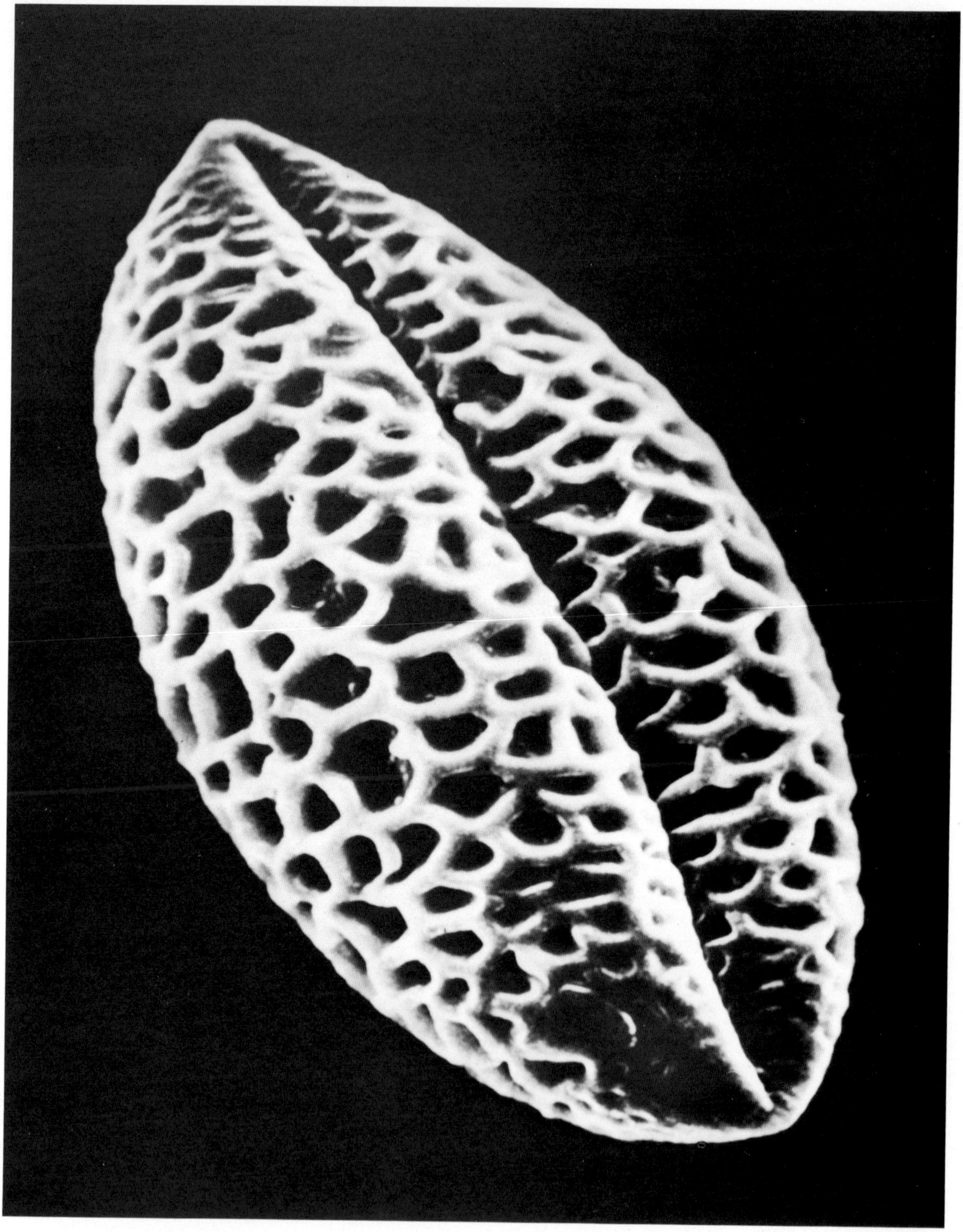

Fig. 2.101 Pollen morphology (79).
Easter Lily (*Lilium longiflorum* Thunb.) [**Liliaceae**] ($\times$ 1600).

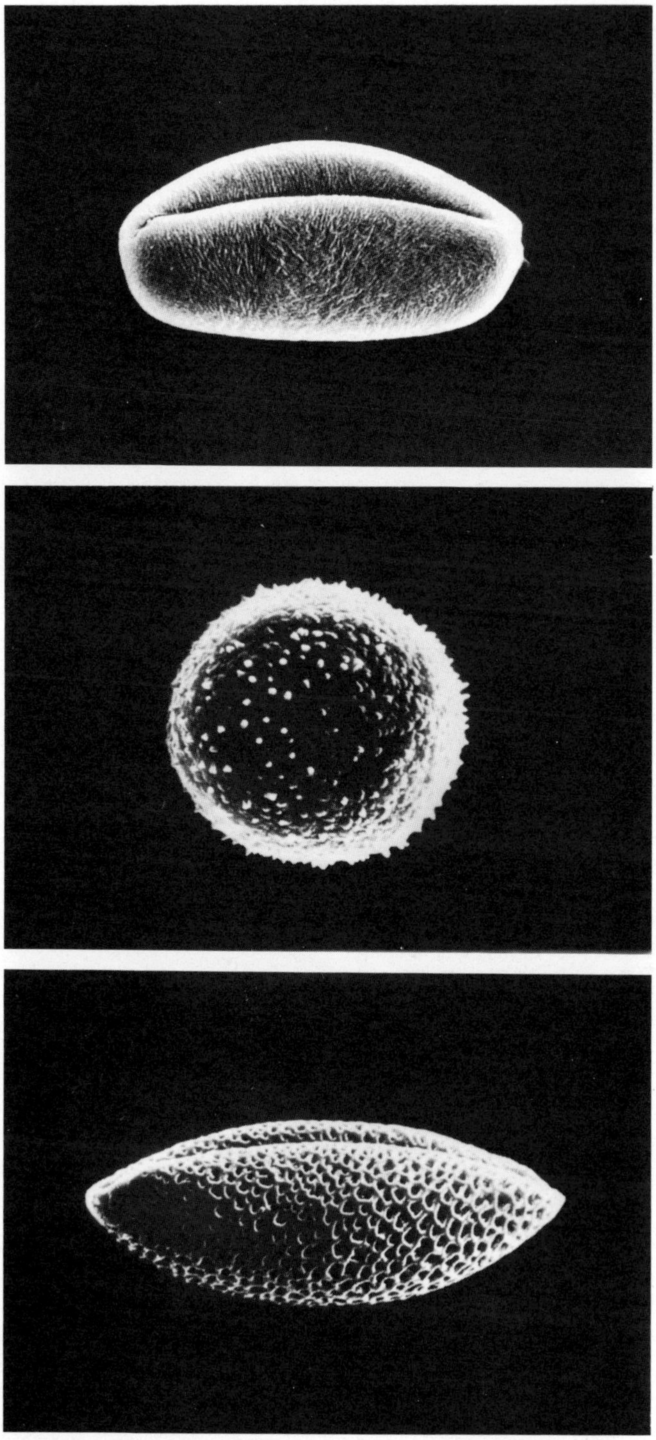

Fig. 2.102 Pollen morphology (80).
Top: *Allium Thunbergii* G. Don [**Amaryllidaceae**] (×1700).
Middle: Greenbrier (*Smilax Sieboldii* Miquel) [**Smilacaceae**] (×1700).
Bottom: Skunk-cabbage (*Symplocarpus nipponicus* Makino) [**Araceae**] (×1700).

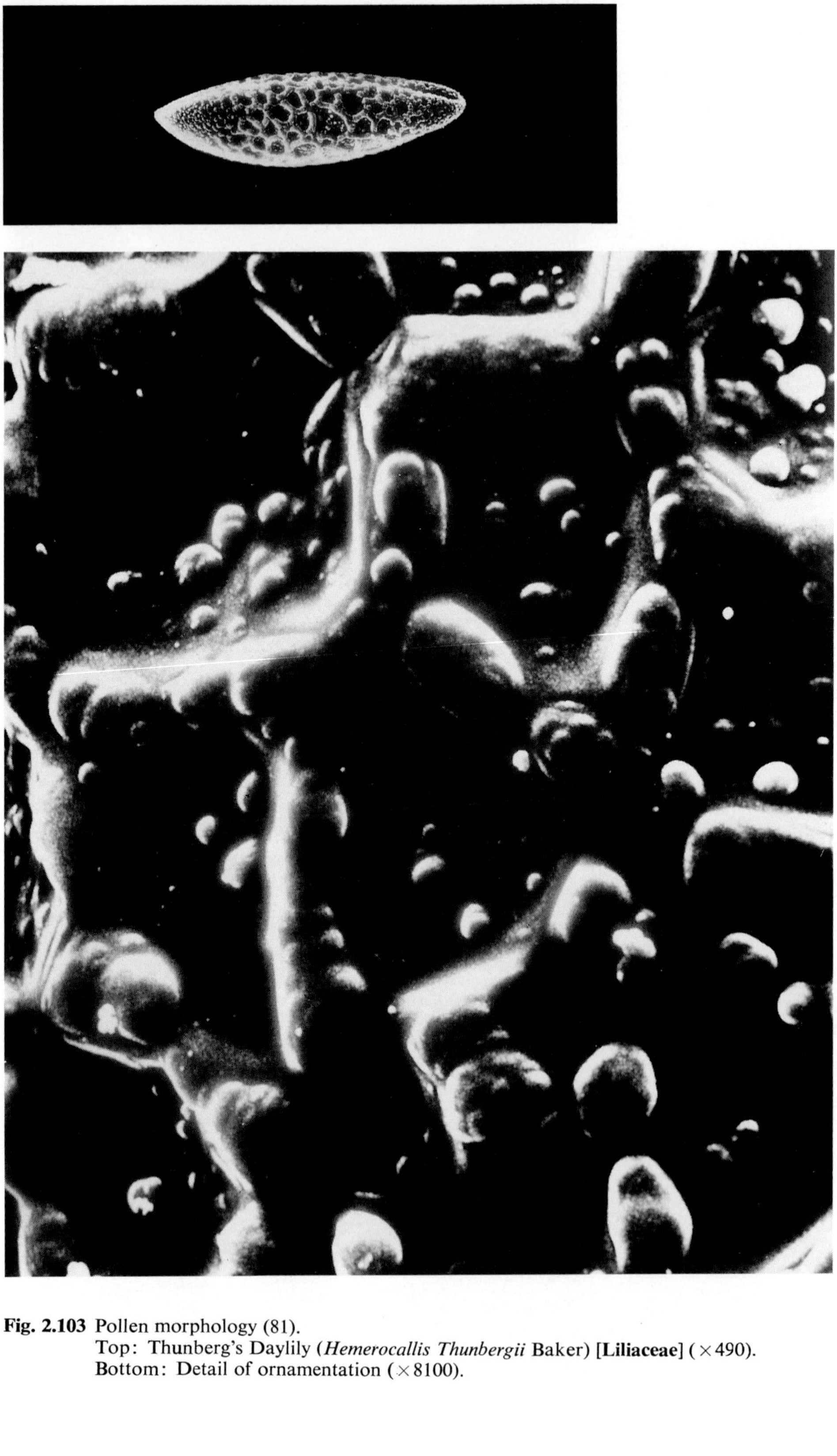

Fig. 2.103 Pollen morphology (81).
Top: Thunberg's Daylily (*Hemerocallis Thunbergii* Baker) [**Liliaceae**] (×490).
Bottom: Detail of ornamentation (×8100).

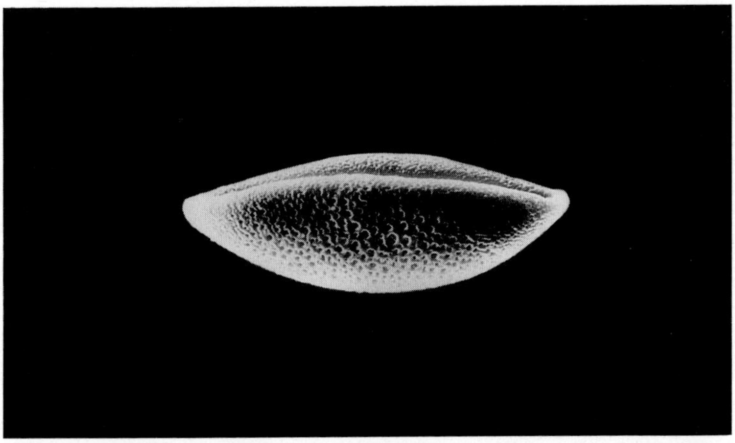

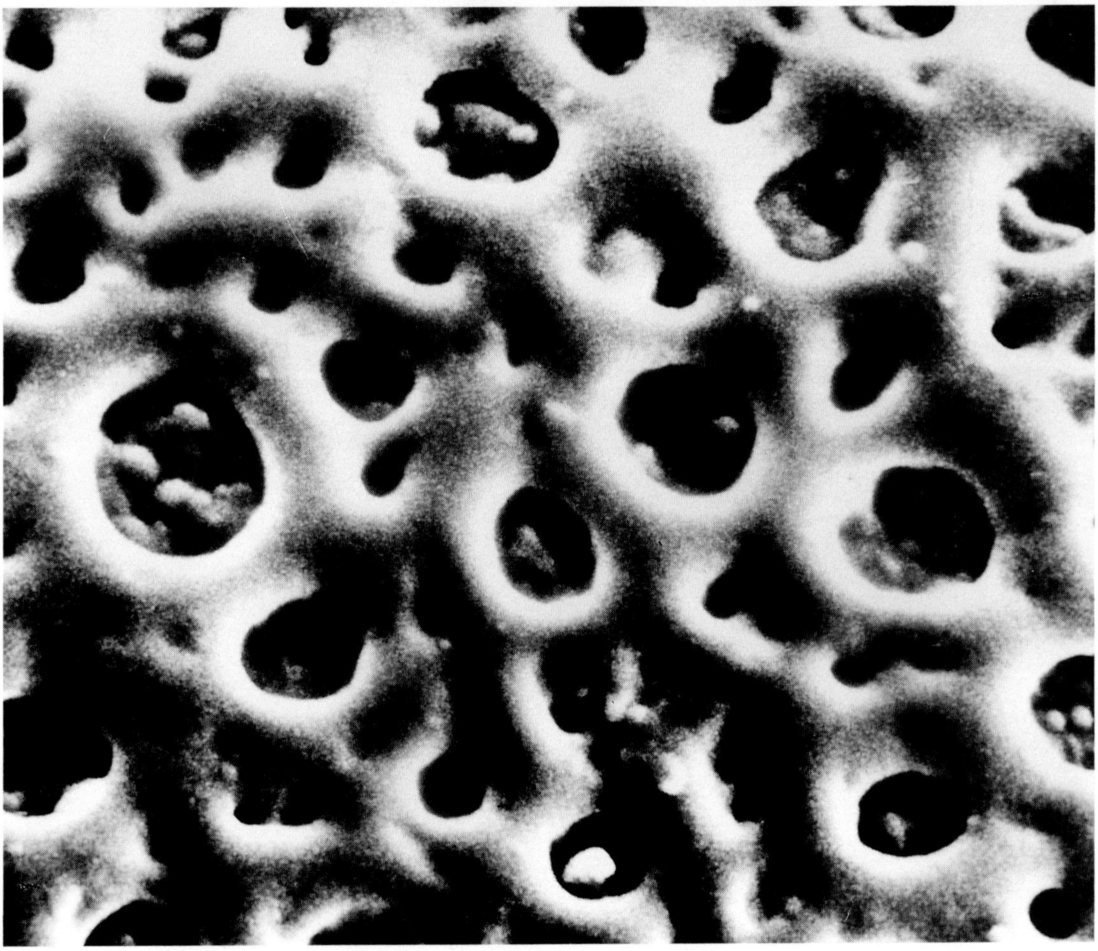

Fig. 2.104 Pollen morphology (82).
Top: *Maianthemum bifolium* DC. var. *dilatatum* Wood [**Liliaceae**] (×1300).
Bottom: Detail of ornamentation (×24400).

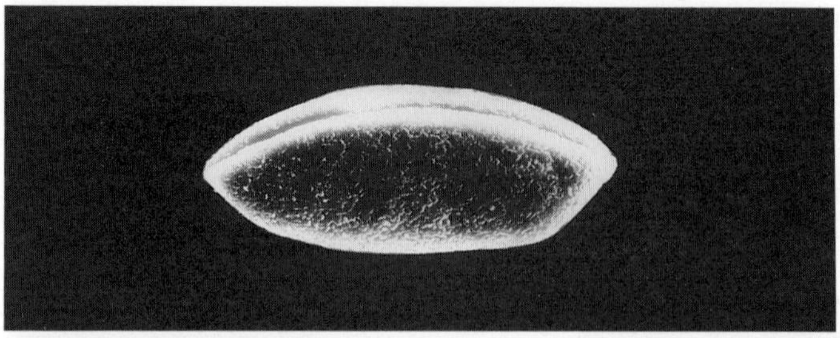

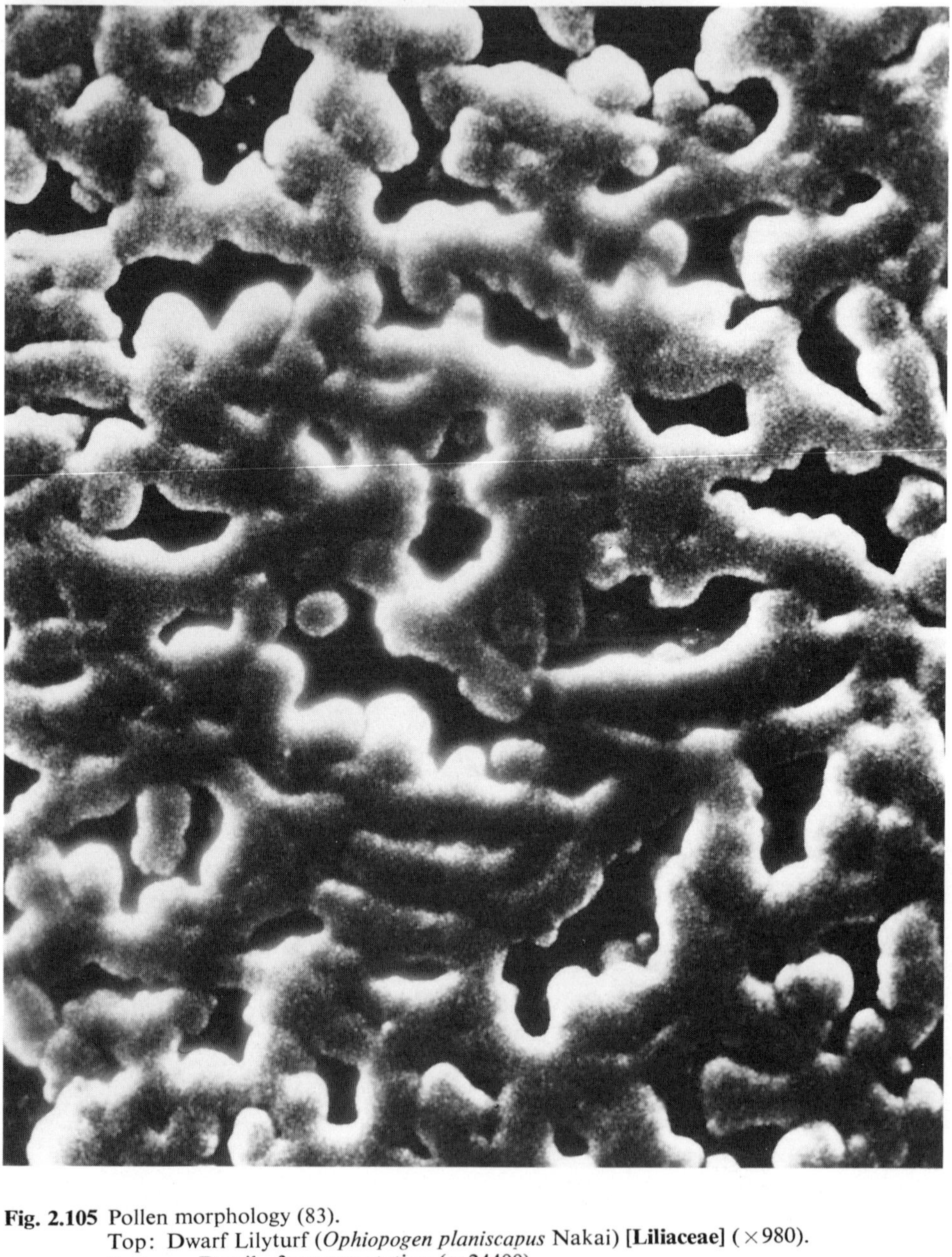

Fig. 2.105 Pollen morphology (83).
Top: Dwarf Lilyturf (*Ophiopogen planiscapus* Nakai) **[Liliaceae]** (×980).
Bottom: Detail of ornamentation (×24400).

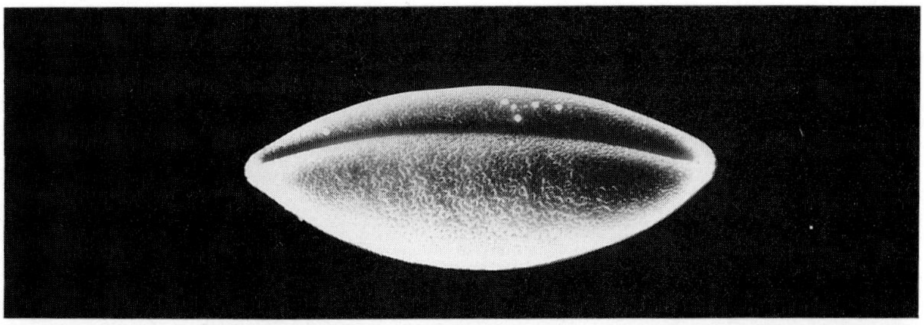

Fig. 2.106 Pollen morphology (84).
Top: Twisted-stalk (*Streptopus streptopoides* Frye et Rigg *var. japonicus* Fassett) [**Liliaceae**] (×980).
Bottom: Detail of ornamentation (×24400).

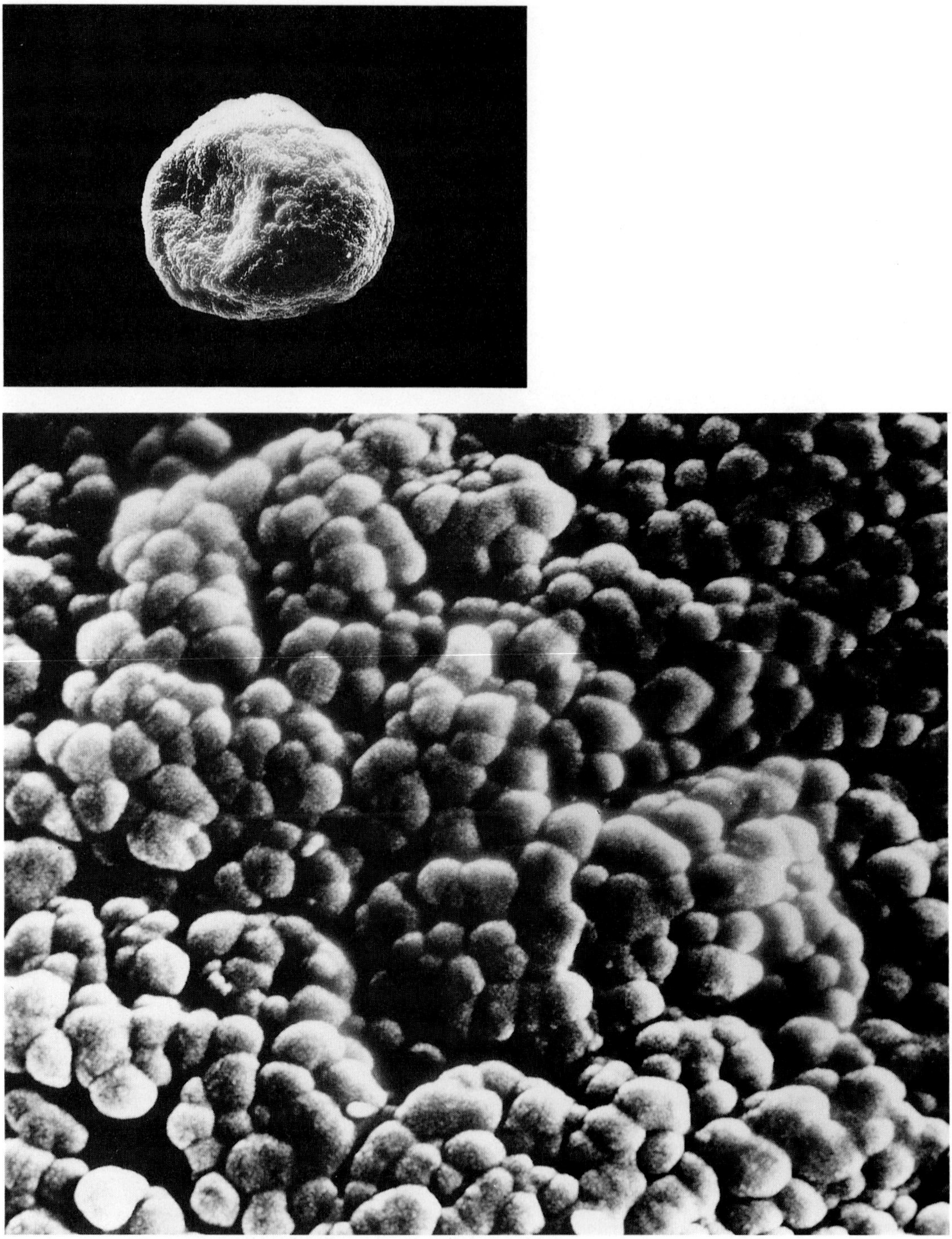

Fig. 2.107 Pollen morphology (85).
Top: Wakerobin (*Trillium Smallii* Maxim.) [**Liliaceae**] (×980).
Bottom: Detail of ornamentation (×16250).

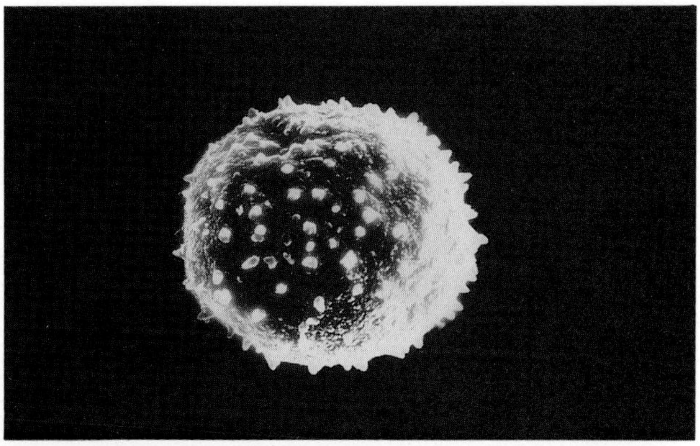

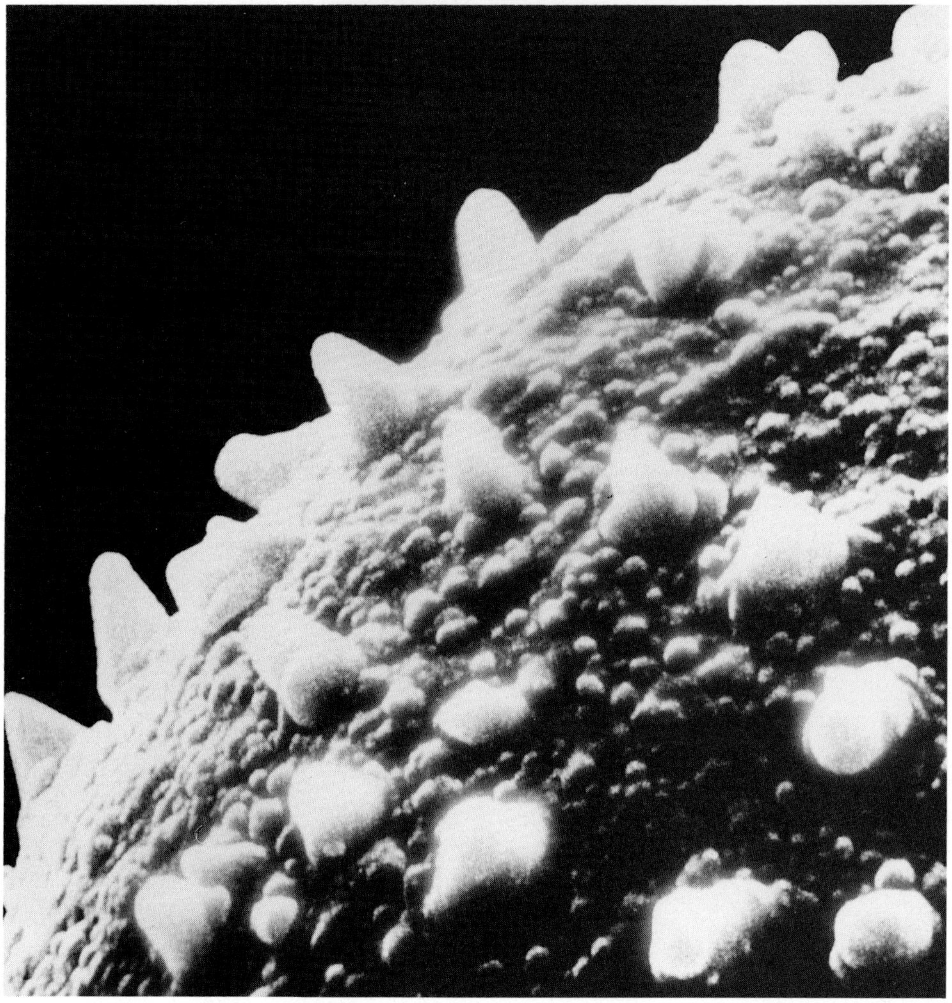

Fig. 2.108. Pollen morphology (86).
Top: Greenbrier (*Smilax nipponica* Miquel) [**Smilacea**] (×1600).
Bottom: Detail of ornamentation (×16250).

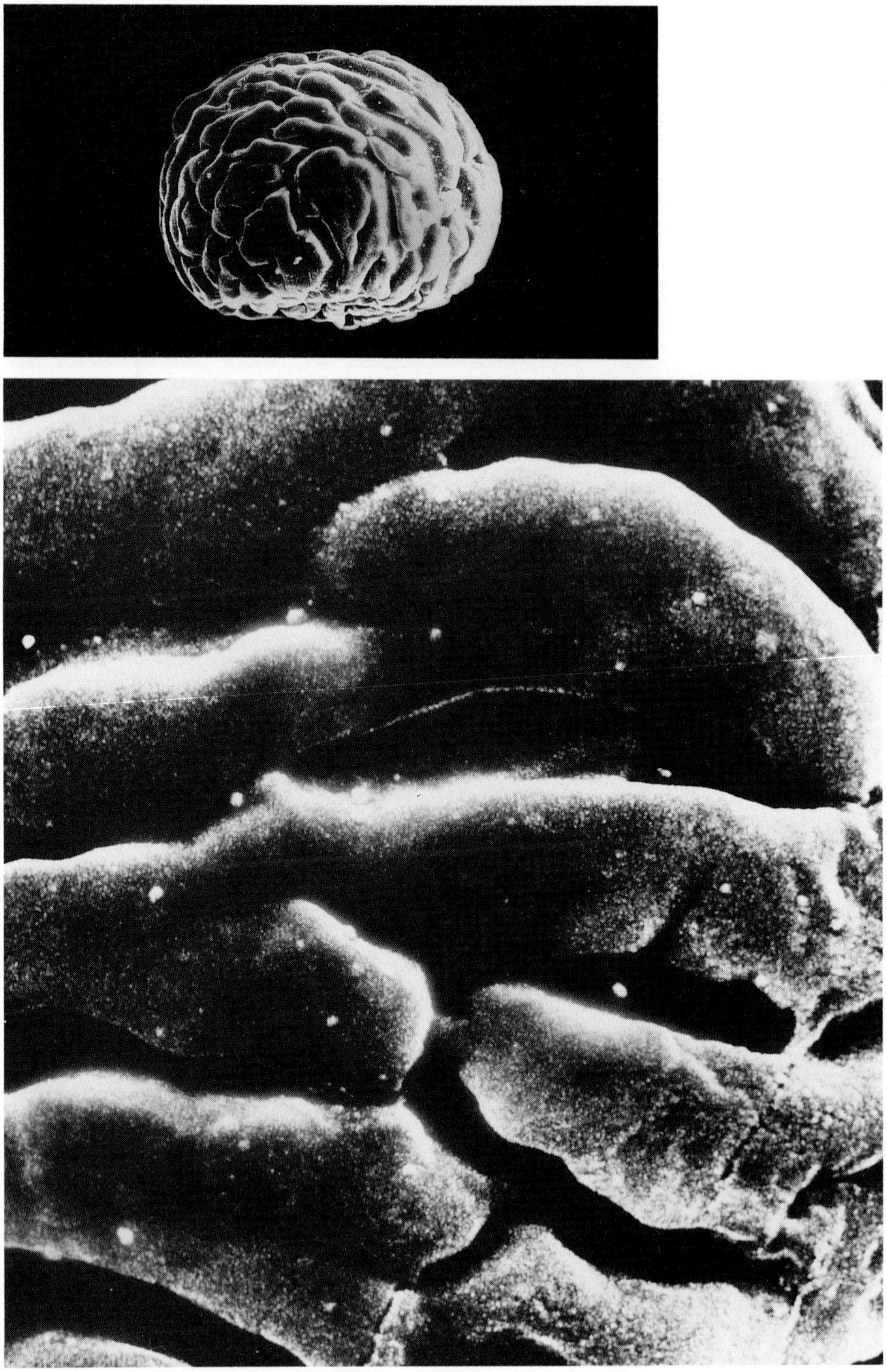

Fig. 2.109 Pollen morphology (87).
Top: Elephant-foot (*Amorphophalus Kanjac* K. Koch.) [**Araceae**] (×1300)
Bottom: Detail of ornamentation (×8600).

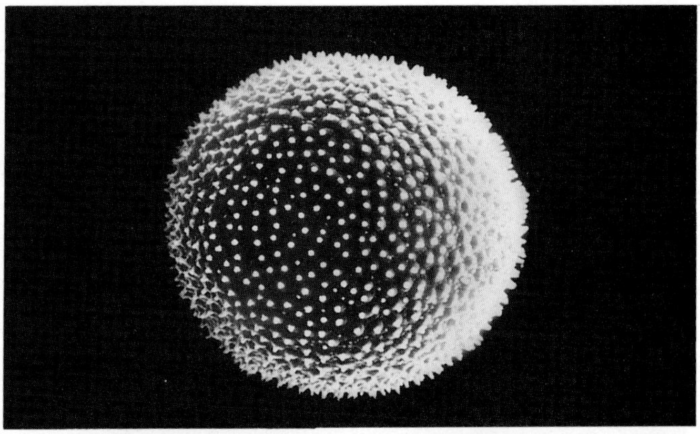

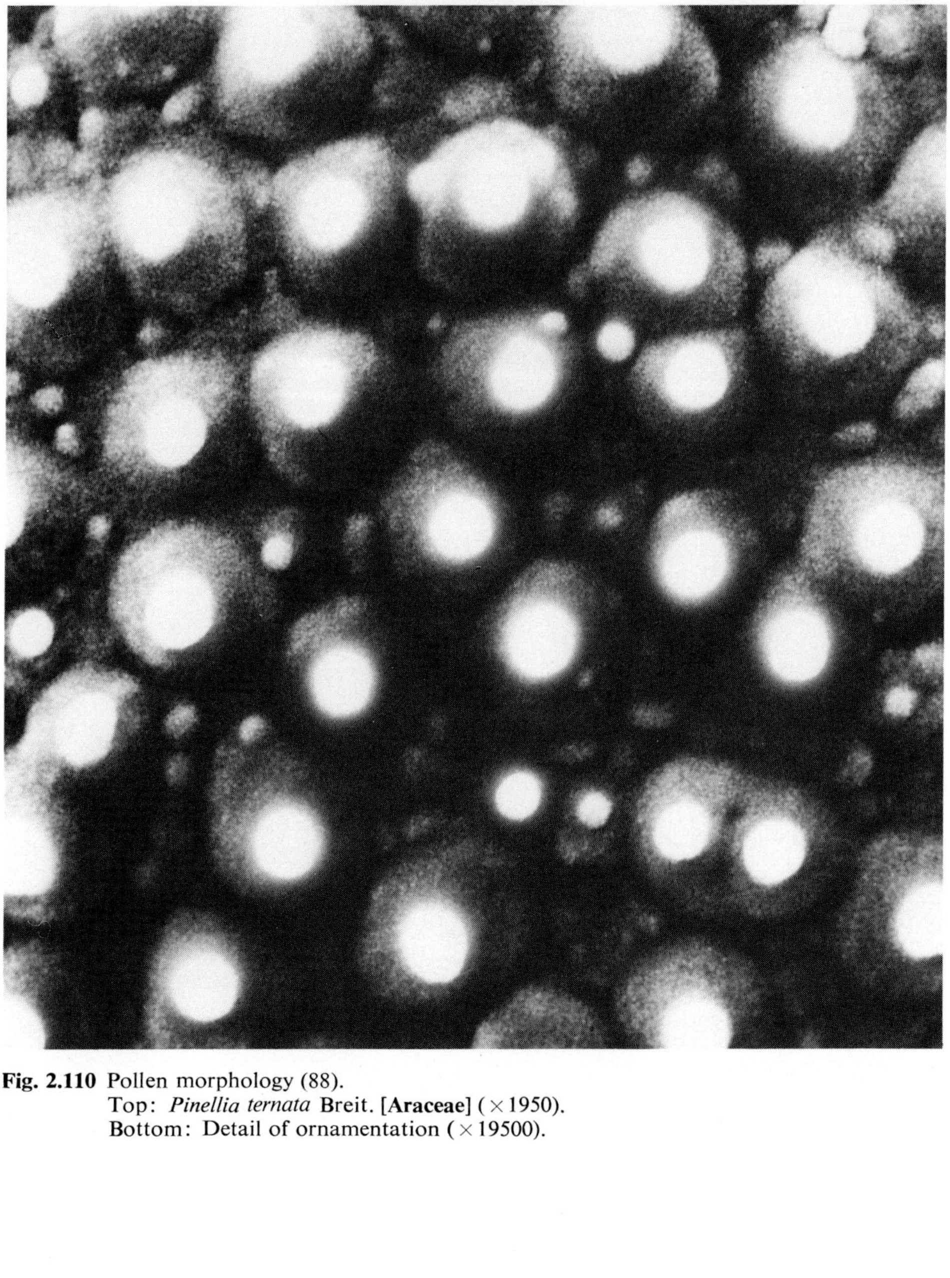

Fig. 2.110 Pollen morphology (88).
Top: *Pinellia ternata* Breit. [**Araceae**] (×1950).
Bottom: Detail of ornamentation (×19500).

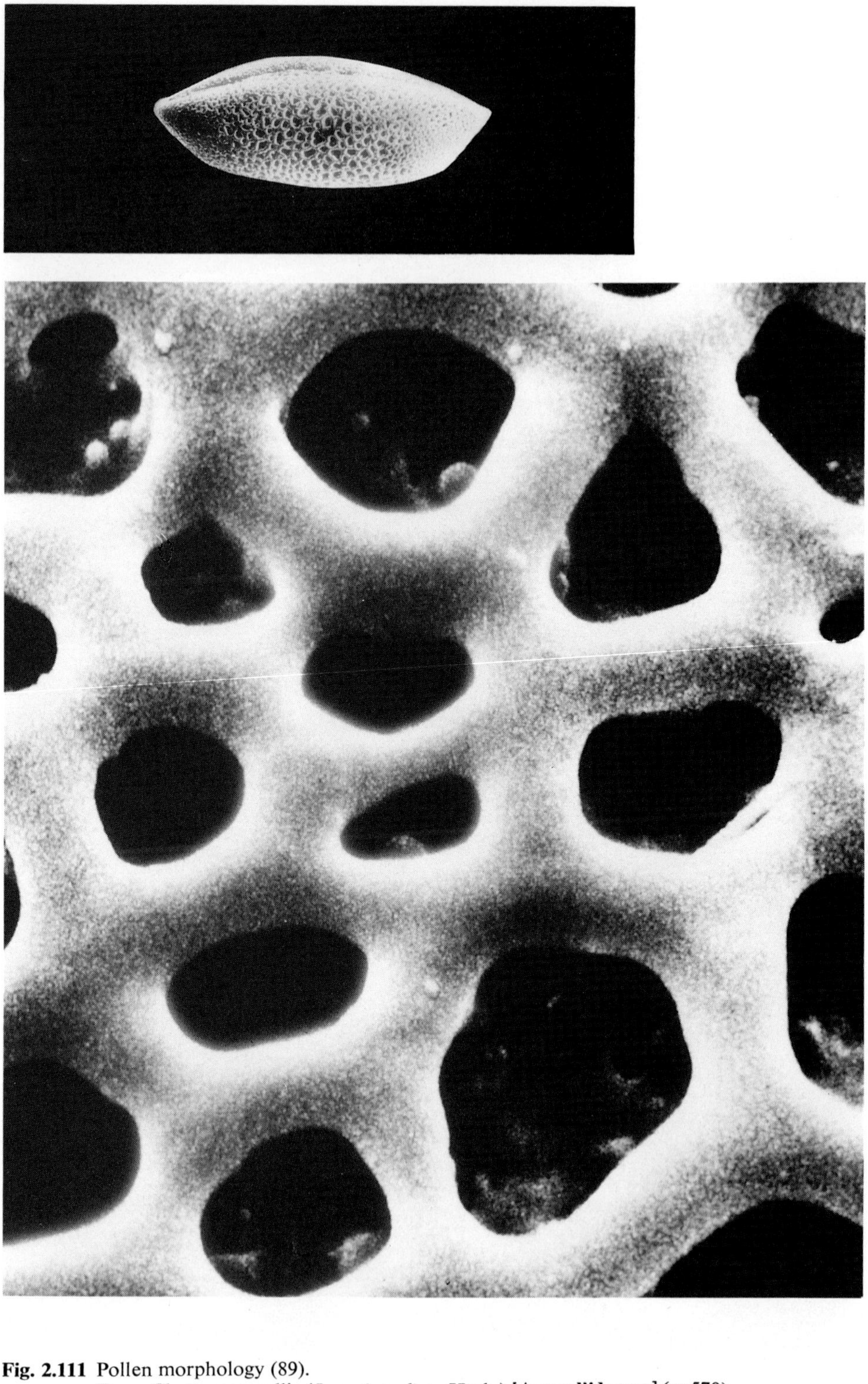

Fig. 2.111 Pollen morphology (89).
Top: Cluster-amaryllis (*Lycoris radiata* Herb.) [**Amaryllidaceae**] (×570).
Bottom: Detail of ornamentation (×13000).

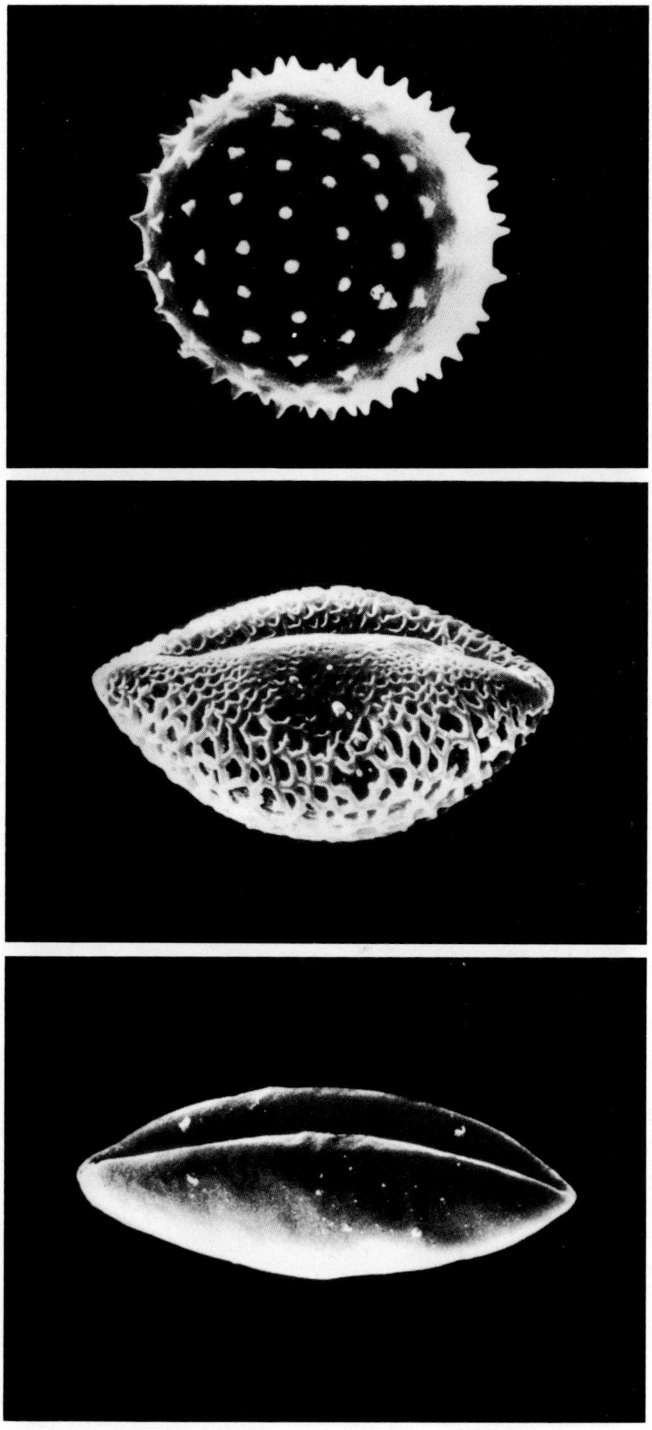

Fig. 2.112 Pollen morphology (90).
Top: Indian-Turnip (*Arisaema japonicum* Blume) [**Araceae**] (×2400).
Middle: Betel Palm (*Areca Catechu* L.) [**Palmae**] (×1700).
Bottom: Bottle Palm (*Mascarena lagenicaulis* L.H. Bailey) [**Palmae**] (×1000).

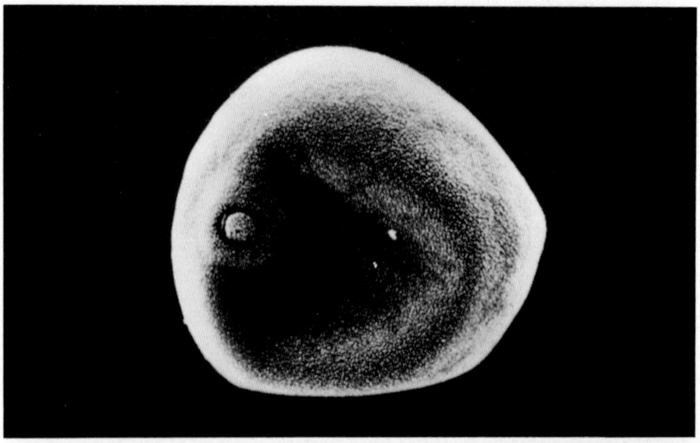

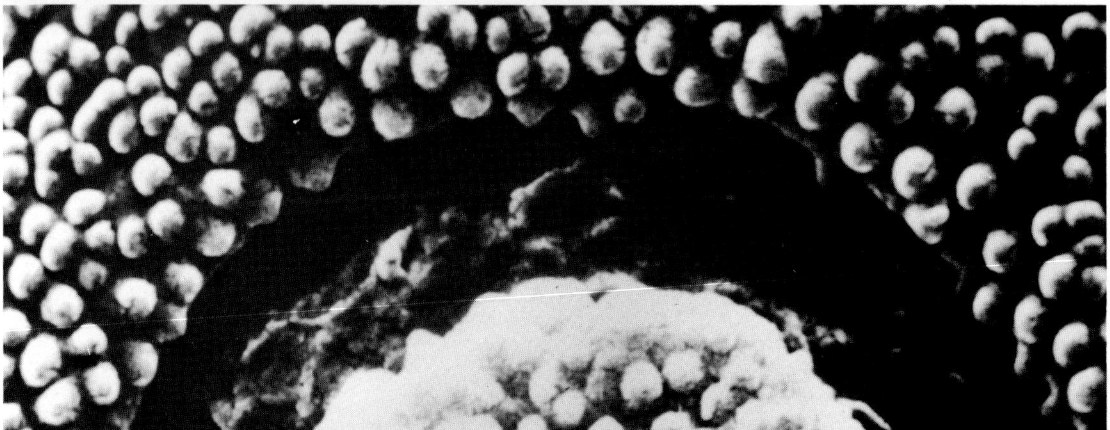

Fig. 2.113 Pollen morphology (91).
Top: Weizein common wheat (*Triticum aestivum* L.) [**Gramineae**] (×800).
Bottom: Aperture with operculum of Weizein common wheat pollen (×20000).

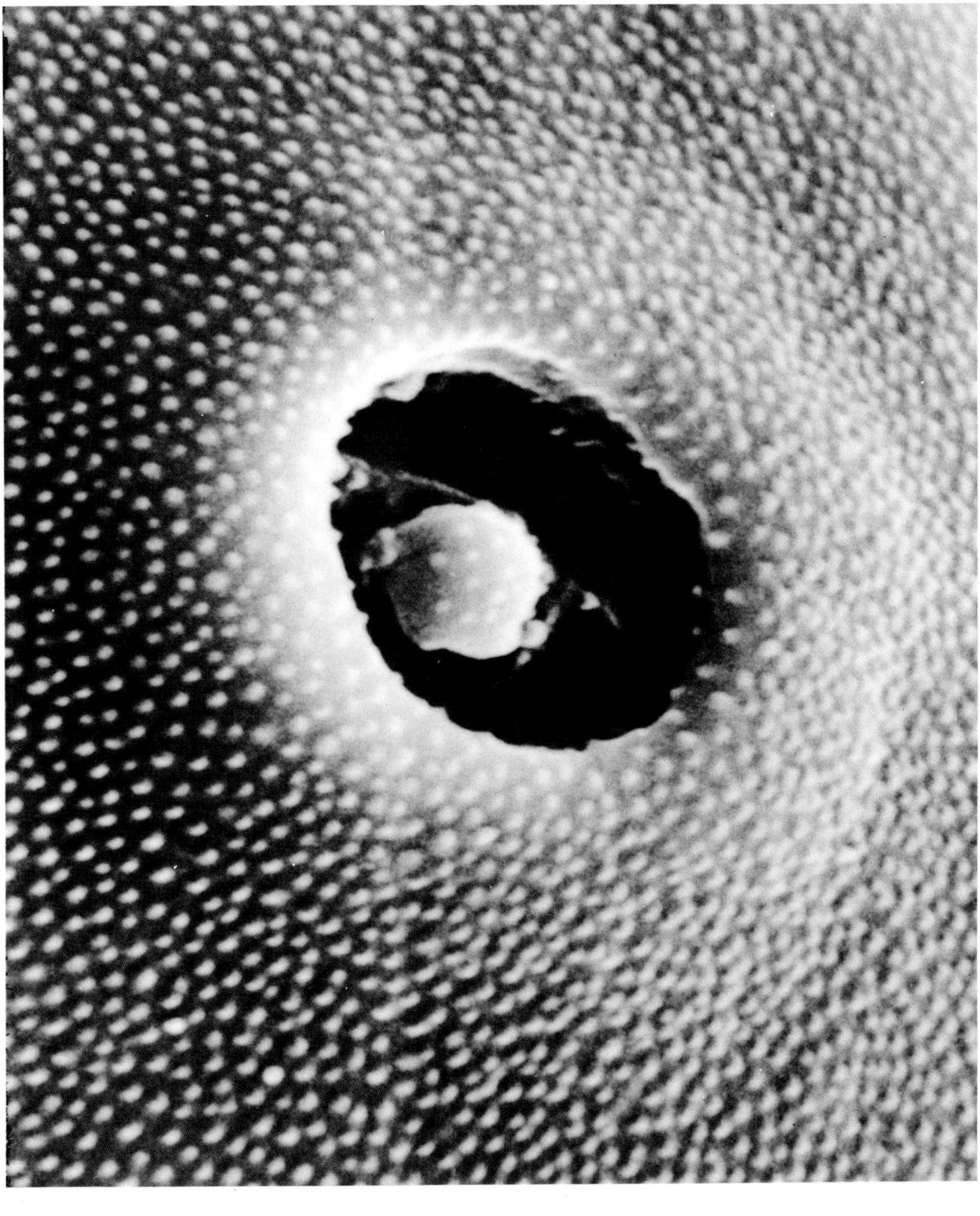

Fig. 2.114 Pollen morphology (92).
Aperture with operculum of Indian corn pollen (*Zea Mays* L.) [**Gramineae**] (×9400).

2.7 Preparation method for scanning electron microscopy

This volume uses scanning electron micrographs to illustrate the many features of pollen grains. Most of the specimens used during the investigations were obtained from plants growing naturally outdoors. Some samples, especially those of tropical plants, were obtained from plants grown in a greenhouse under the supervision of the Parks. In all cases, pollen samples were obtained only from the most mature specimens available.

Prior to viewing a sample in SEM, it is necessary to dry and surface-coat the specimen with metal. Most of the samples used were air dried at room temperature for about 30 minutes. Some samples were prepared by critical point drying. After drying, the sample was attached to stubs by double-stick tape and placed in a vacuum-evaporator with a rotating stage for coating carbon and gold. The samples were examined and photographed in a JEOL JEM-15 scanning electron microscope operated at 15 kV. Size calculations were approximate because of the variability of specimens with regard to beam angles inherent to SEM. In the morphological analysis of pollen, each grain should be observed from four directions, which are sometimes compared to the directions of the globe, *i.e.* polar views from north and south, an equatorial view and a side view at a 90-degree angle from the equatorial view. Most of the micrographs in this volume have been taken from the equatorial view because it reveals characteristic surface morphologies including exine structures and the locations of apertures.

3. Pollination and Pollen Growth

Pollen have various shapes depending on the species and have species specific patterns on their surface, as described in chap. 2. The biological significance of these sophisticated surface patterns remains a mystery. These structures may function for attachment to insect legs. The designs, however, are much too delicate for the size of insect legs or hairs (see Fig. 3.6). Furthermore, there is no reason for each species to have a specific pattern for this purpose. At any rate, when flowers open, pollen grains are released from the anthers onto stamens to pollinate pistils.

3.1 Flowers open in order to receive pollen

Many beautiful flowers bloom on mountain tops or in deep jungles where men seldom go. Needless to say, flowers do not exist for human beings but bloom to scatter pollen and attract insects in order to achieve fertilization.

In early morning, as soon as the flowers of the rose moss open, pollen grains migrate onto the stigma, and after fertilization, the flowers close in the afternoon. It takes about 3.5 hours for the flowers of the rose moss to close completely after pollination. However, when a flower is prevented from receiving pollen by isolation or emasculation (removal of anthers), it remains open longer (Fig. 3.1).

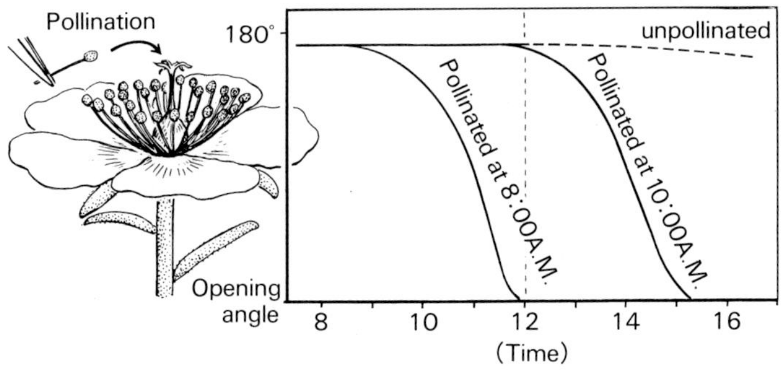

Fig. 3.1 Effect of pollination on flower opening in rose moss (*Portulaca grandiflora*). Flower closes soon after the pistil receives pollen grains.

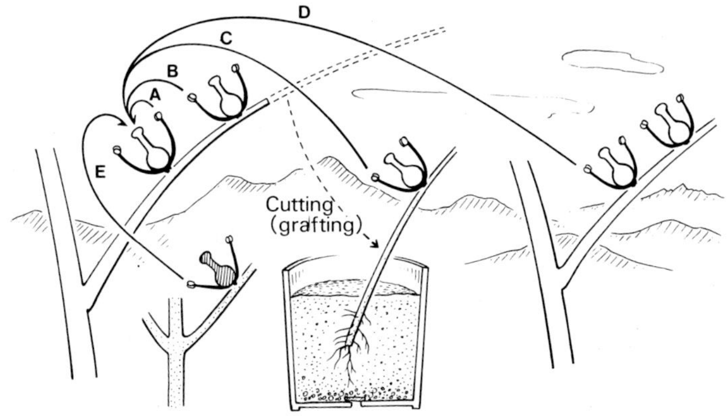

Fig. 3.2 Variation of pollination. A: self-pollination within a flower. B: self-pollination within a plant (geitonogamy). C: self-pollination between plants (semi-geitonogamy). D: out-corssing pollination. E: hybrid-crossing pollination.

Fig. 3.3 Opening and closing of wheat florets. Flowering florets extending white pistils, indicated by arrows, close completely one hour after pollination.

Glumes of wheat floret also begin to close twenty minutes after pollination (Fig. 3.3), but stay open for about two weeks if the anthers in the florets are taken off before maturation and isolated from out-crossing. These phenomena clearly indicate that flowers open in order to receive pollen.

3.2 Various types of pollination

There are various types of pollination, the process by which pollen grains attach to the stigma (Fig. 3.2). For example, self-pollination means not only pollination within a single flower, but pollination within a single plant (geitonogamy) as well as between the original plant and a cutting clone (semi-geitonogamy). These are genetically inbreeding crosses.

Out-pollination is pollination by which the stigma receives pollen from another plant of the same species (intra-specific fertilization). Hybrid pollination occurs if pollen from one species attaches to the stigma of the other species to bear hybrid fruits, as in the case of pollen grains of peach migrating onto the stigma of cherry blossoms (inter-specific fertilization).

3.3 Migration of pollen grains

Pollen grains are not equipped to move by themselves, but migrate with the help of animals, wind, or water. Insects are the major pollinators among animals. Flowers pollinated by insects are called, entomophilous flowers (Fig. 3.4).

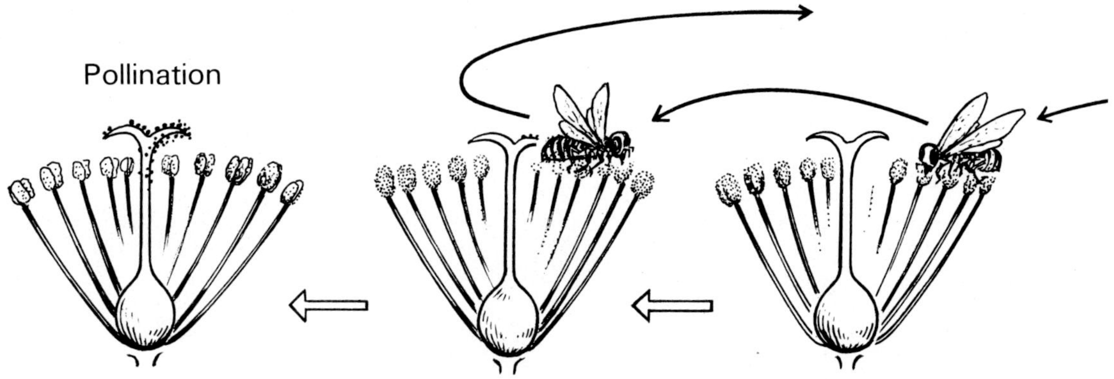

Fig. 3.4 Entomophilous (or ornithophilous) pollination.

There are many trees whose pollen migration depends on wind (anemophilous flowers). The structures of these flowers are relatively simple, sometimes without petals and smell which are required for attracting insects, but they produce relatively small and light pollen grains. In Fig. 3.5, migration speed (cm/sec) of pollen grains in various species are presented under no-wind conditions. Since pollen grains easily float in air, they can be airborne to reach the pistils of other plants.

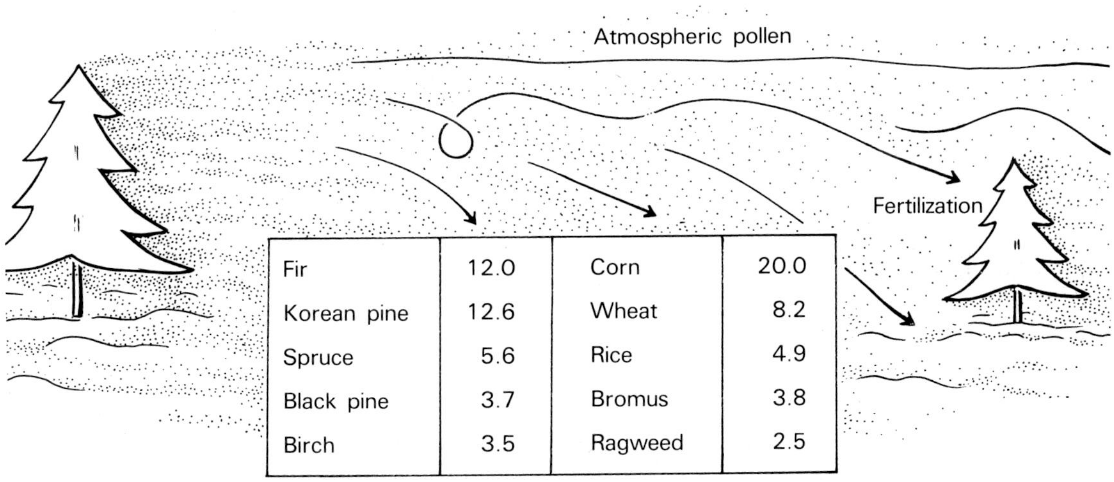

Fir	12.0	Corn	20.0
Korean pine	12.6	Wheat	8.2
Spruce	5.6	Rice	4.9
Black pine	3.7	Bromus	3.8
Birch	3.5	Ragweed	2.5

Fig. 3.5 Migration speed (cm/sec under no wind) of pollen grains in anemophilous plants.

Fig. 3.6 Pollen grains of strawberry on the tibia of an aphis larva. Note the many pegs of the tibia and large clumps of pollen. ×1300.

3.4 Honeybees and pollen

Flowers attract not only butterflies and honeybees, but many small animals as well. Fig. 3.6 is a SEM micrograph of a larva of aphis carrying strawberry pollen grains on its tibia, demonstrating that aphides function as pollinators of plants.

Honeybees collect nectar from flowers and accumulate honey with pollen grains in the beehive (Fig. 3.9). They imbibe honey into honey sacs and carry a mass of pollen grains called pollen load. Both are nutrient sources for bees. Under microscopic observation, it was noted that one milliliter of honey contains about twenty thousand of pollen grains (Figs. 3.7, 3.8). Since honey is rich in various substances including amino acids, proteins, enzymes and vitamins secreted from pollen and bees, it is not only food for bees, but a rich nutritious food for human beings, with concentrated nectar mixed with the above substances. In return for nectar to make honey from plants, bees serve as pollinators for the flowers. This is one example of a perfect give-and-take relationship in nature.

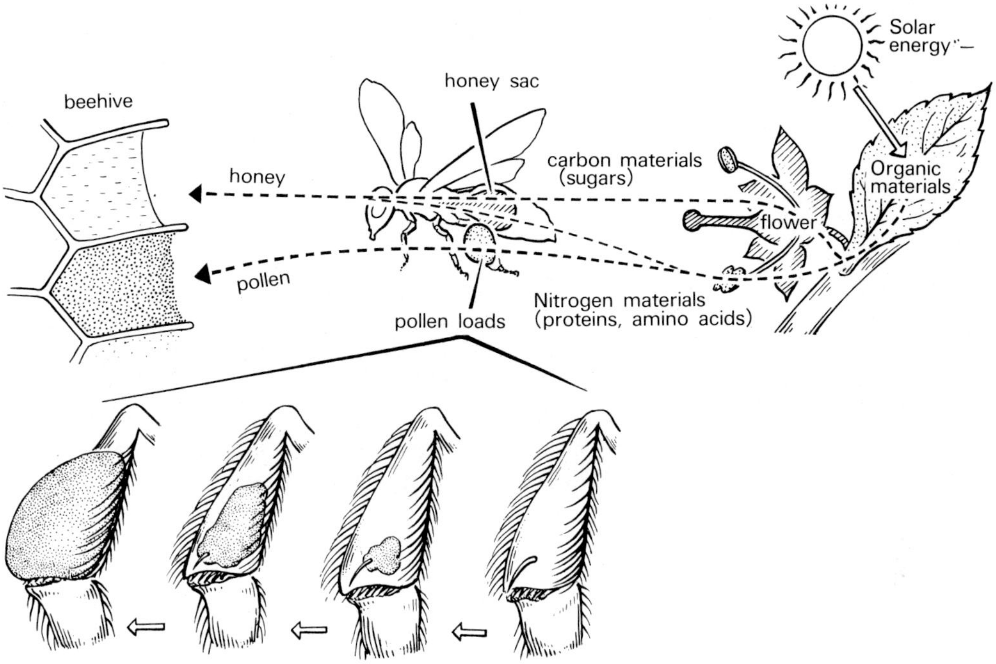

Fig. 3.7 Honeybees storing energy resources produced by photosynthesis. A honeybee makes two pollen loads in pollen baskets of the hind legs.

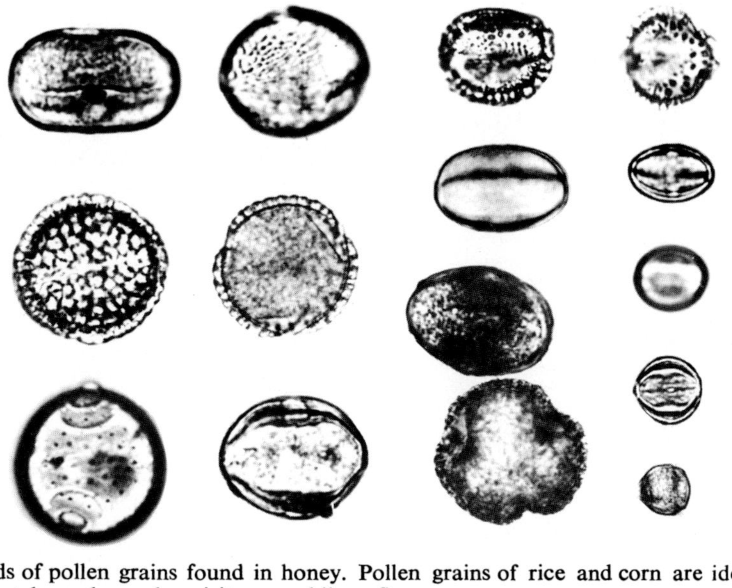

Fig. 3.8 Various kinds of pollen grains found in honey. Pollen grains of rice and corn are identified in honey, indicating that honeybees also visit nemophilous flowers.

Fig. 3.9 Pollen loads (white parts) stored by honeybees in their hive.

3.5 Pollen loads

Pollen load is the mass of pollen grains made by bees (Fig. 3.7). The size of one load is about 3 × 1.5 mm, weighing 2–10 mg. Bees carry larger loads in the morning than in the afternoon.

Pollen is rich in sugars, starches, amino acids, vitamins, and hormones, so it is natural for man to utilize these components for food and medicine. On the market these days, we find many pollen-containing products, including various types of food and cosmetics. Pollen has become very popular recently as a health food. Pollen loads are collected at the entrance of beehives. Figs. 3.10 and 3.11 show the method for collecting pollen loads from bees' legs when they pass through holes set up at the entrance of the beehive box.

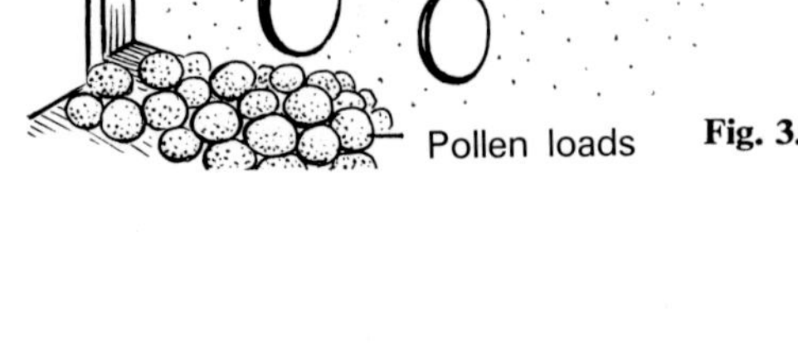

Fig. 3.10 Pollen load collecter used in bee yards.

Fig. 3.11 Pollen load collecter. Pollen loads are removed from bees' legs whenever honeybees pass through the small holes to their hive.

3.6 Movement of the stamen

When the stamens of barberry (*Berberis thunbergii*) are touched by fingers, they bend inwards (Fig. 3.12). The stamens of rose moss also curve against the pushing force when pushed with the tip of a pencil (Fig. 3.12). This movement is useful for spreading pollen grains over the insect's

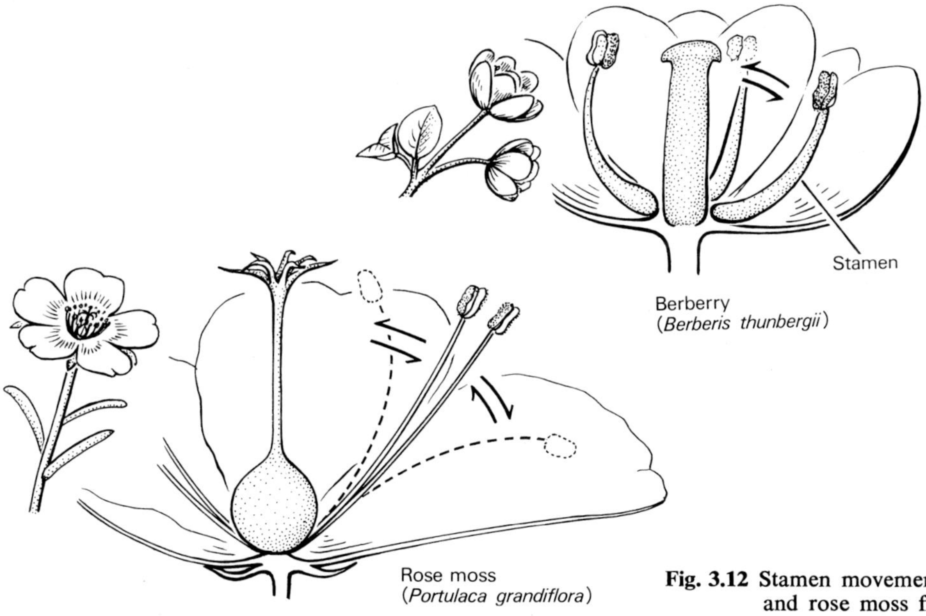

Fig. 3.12 Stamen movement of berberry and rose moss flowers.

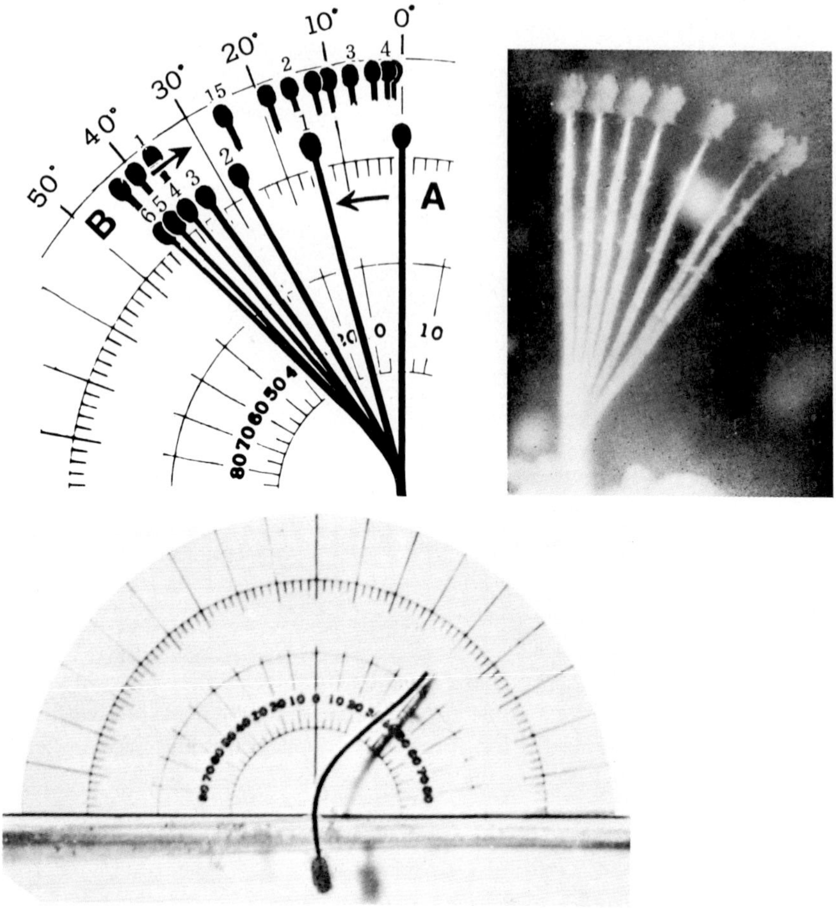

Fig. 3.13 Process of stamen movement of rose moss. Stamens bend for pushing power then rebound. Top left: when touched from left to right, it bends 40 degrees to the left within 6 seconds (A), then returns to its original position in 4 minutes. Top right: Sequence photographs of bending to the right. Bottom: Stamen movement is retained even when excised and stimulated.

body. Fig. 3.13 indicates the direction of movement when the excised stamens of rose moss are placed on a glass slide and stimulated. It was found that the stamen lost its bending ability against stimulation if the flower was treated with ether; the ability was recovered when ether was removed.

3.7 Dichogamous flowers (how inbreeding is avoided)

Human society prohibits inbreeding by cultural custom and/or laws to avoid the disadvantages of increased genetic homozygosity. If pollination occurs within the flower of a monoecious plant, inbred offspring are produced. There exist various mechanisms in plant society to avoid such inbreeding.

Dichogamy is one of these systems in which the male organs mature at a different time than the female organs of the same flower in order to avoid self-pollination (Fig. 3.14). For example, the pistil matures earlier than the stamens in the flower of lawn grass (*Zoysia japonica*). On the contrary, the stamens of the balloon flower (*Platycodon grandiflorum*) mature and scatter pollen grains in the air 1–2 days before maturation of the pistil. The former type is designated as protogyny and the latter is called protandry. Pistils of these flowers receive pollen from other plant individuals, while their own pollen are utilized to fertilize the pistils of other flowers. The protogyny species includes lawn grass (*Zoysia japonica*), plantain (*Plantago asiatica*), bithwort (*Aristolochia debilis*) and lily magnolia (*Magnolia liliflora*). Protandry plants include balloon flower (*Platycodon gradiflorum*), bellflower (*Campanula punctata*), touch-me-not (*Impatiens balsamina*), mother-of-thousands (*Saxifnaga stelonifera*), and others (Figs. 3.15, 3.16, 3.17).

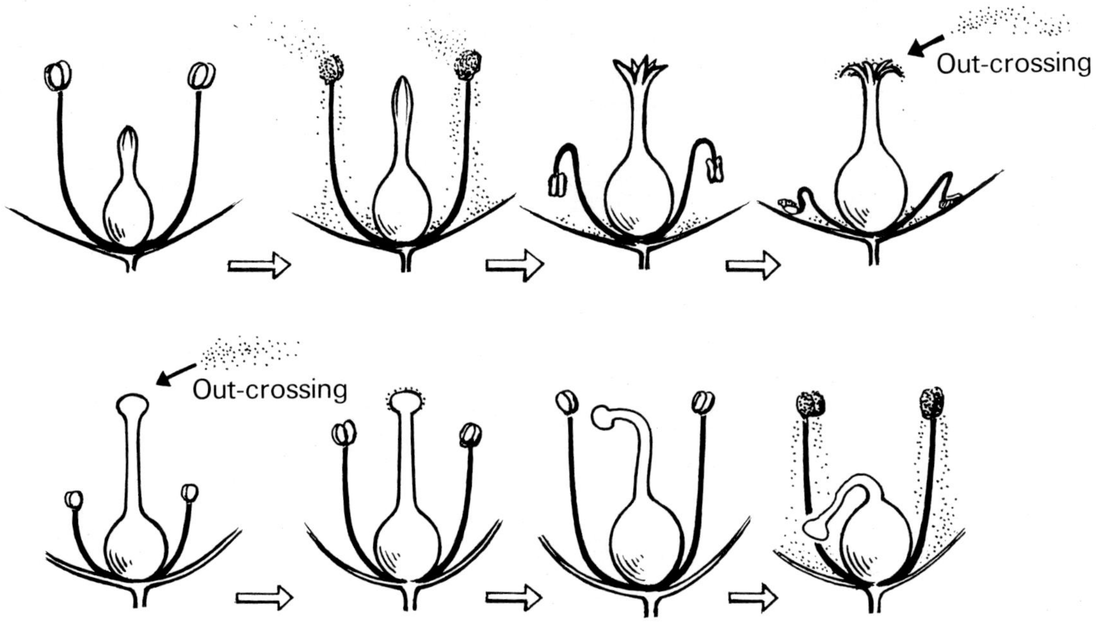

Fig. 3.14 Dichogamous flowering advantageous for out-crossing. Top: Protogyny type. Bottom: Protandry type.

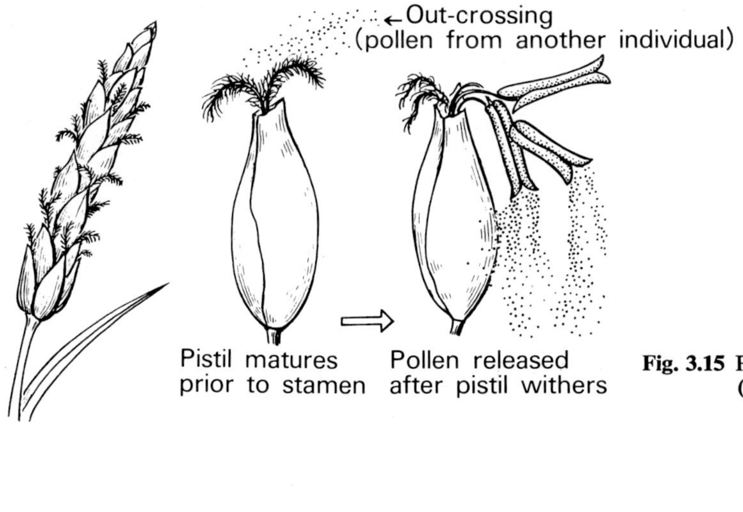

Fig. 3.15 Protandry system in berberry flower (*Berberis thunbergii*).

Fig. 3.16 Protogyny in plantain (*Plantago asiatica*) and lawn grass (*Zoysia japonica*).
Left: Pistil develops prior (*left*) to anthesis (*right*). Right: Pistil matures first (*top*), followed by anthers (*bottom*).

Fig. 3.17 Protandry of balloon flower.
Left: flowering time. *Right:* two days after initiation of flowering.

There are plants which have systems other than dichogamy for avoiding inbreeding. For example, weeping forsythia (*Forsythia suspensa*) and trumpet lily (*Lilium longiflorum*) have a self-incompatible mechanism by which pollen can not germinate or pollen tubes stop growing on the stigma when they are genetically identical (see Fig. 3.28).

3.8 Stigma reactions

Stigma reactions are physiological changes which occur as a result of pollination. For example, when pollen grains of wheat (*Triticum aestivum*) or rice (*Oryza sativa*) are placed on a stigma of wheat, the nuclei of stigma cells of wheat become easily stainable (Fig. 3.18). A similar change can be observed in the stigma tissue of rice upon pollination with pollen of rice (Fig. 3.19) or wheat. This indicates that pollen grains stimulate stigma cells in order to be activated.

Soon after pollinated pollen initiate germination on the stigma, the stigma starts to wither and rejects further pollen germination. This is another example of stigma reaction.

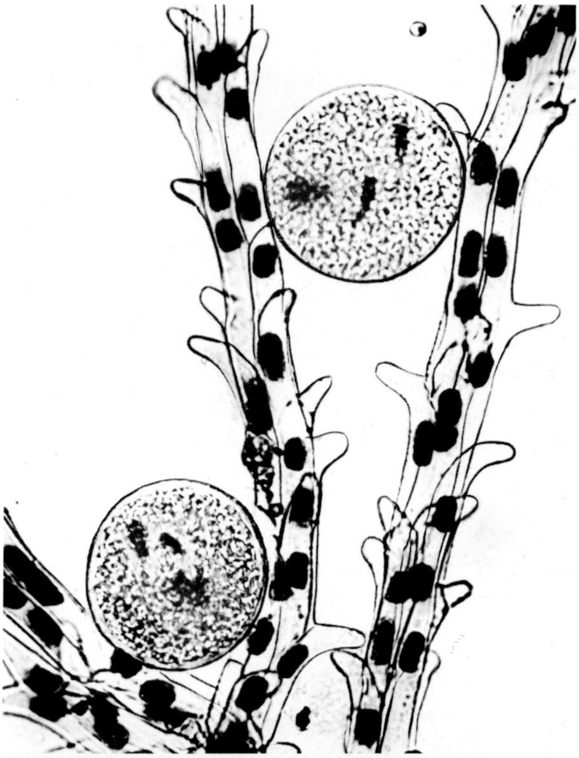

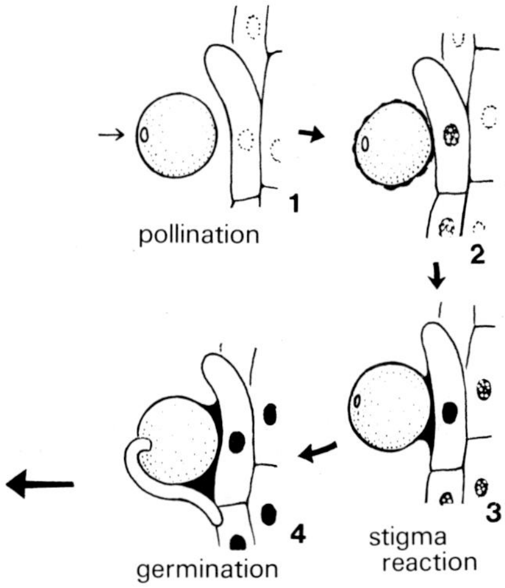

Fig. 3.18 Stigma reaction in pollination of wheat (*Triticum aestivum*). The nuclei of stigma cells become stainable with carmine dye starting from the point to which the pollen grain attaches itself.

Fig. 3.19 Pistil and stamens of rice (*Oriza sativa*). The pistil consists of two branching stigma; and stamen has six anthers.

3.9 Pollen growth

Pollen grains on the stigma absorb water from pistil tissue and pollen tubes start to elongate. In rice, wheat, or rose moss, the stigma is hairy, and pollen tubes elongate along the cell layers of the stigma to enter the pistil (Fig. 3.20). The stigma of azalea (*Rhododendron*) is not hairy, and pollen tubes grow along the surface seeking a crack through which to enter (Figs. 3.22, 3.23, 3.24).

On the surface or inside the stigma, mucilage containing sugars, amino acids and growth hormones is secreted due to stigma reaction. Pollen tubes elongate in the mucilage, obtaining energy and nutrition for growth of hundreds and thousands times grain size, to reach the ovules in the ovaries located at the basal portion of the pistil (Fig. 3.26).

The time required from termination of pollen germination to tube arrival to the ovules varies among species. For example, in touch-me-not and rose moss the process takes one to two hours while in azalea it requires 24 hours. Two to three days are needed for the lily to complete the process.

Fig. 3.20 Germinating pollen on stigma of rose moss. Pollen grains are stained with iodine, showing elongating tubes inside the pistil tissue.

Fig. 3.21 A pollen grain of rose moss (*Portulaca grandiflora*). ×600

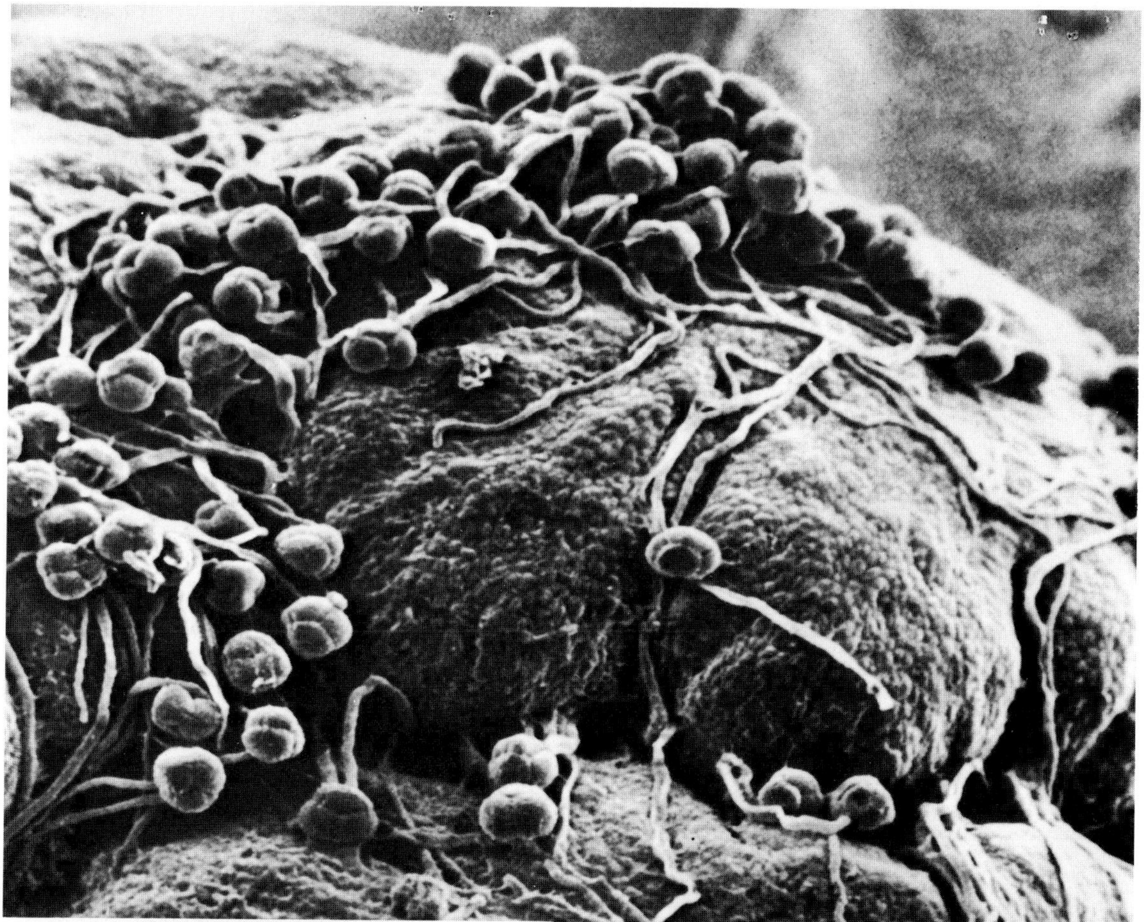

Fig. 3.22 Surface view of the stigma of azalea (*Rhododendron pulchrum*) 36 hours after pollination showing pollen tubes emerging from colpi of grain. ×200.

Fig. 3.23 Longitudinal section of a pollinated stigma of azalea showing part of the stigma surface and a portion of the top of the style. Note that tubes are growing in the stylar canal. ×50.

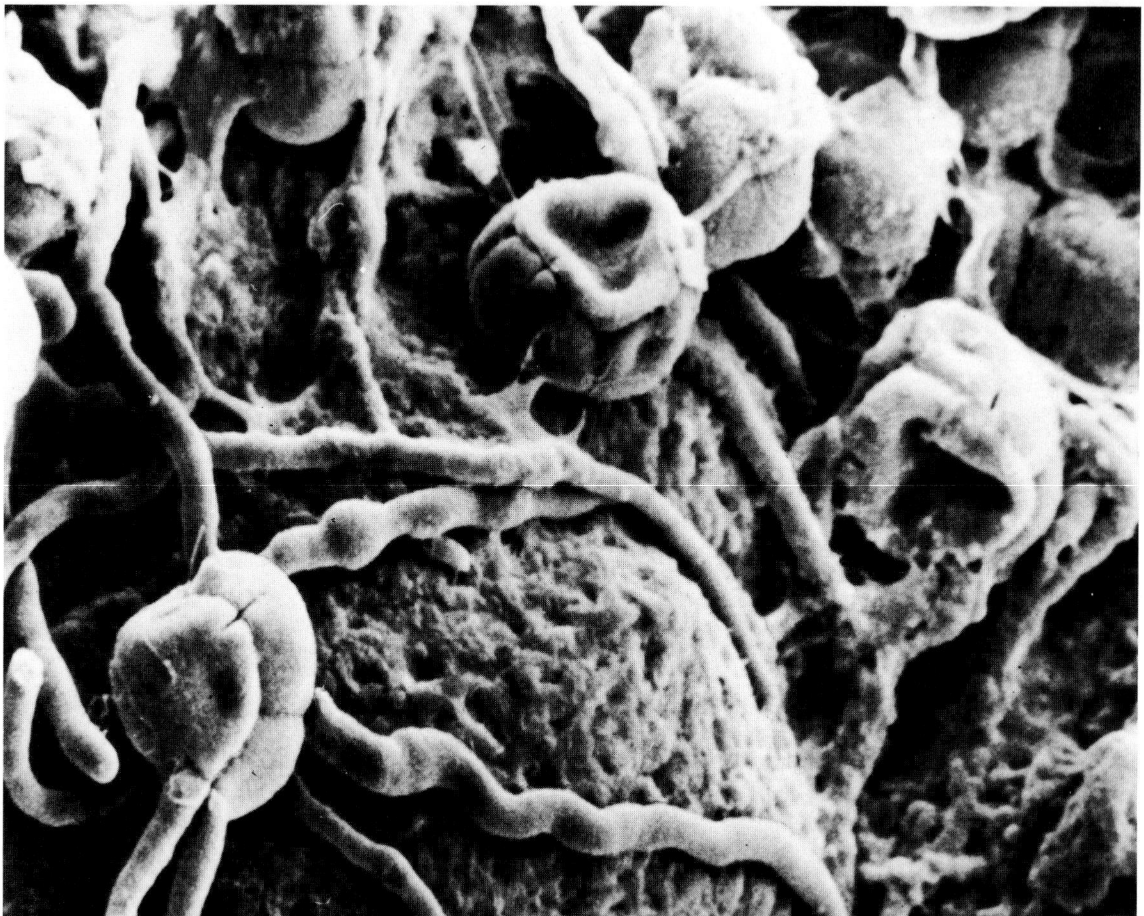

Fig. 3.24 Detail of stigma surface and pollen tubes shown in Fig. 3.22. The tube tip growing through the stigmatic structure. ×800.

Fig. 3.25 SEM micrograph of growing pollen tubes at the stigmatic suture. After pollination, exudation of the interstitial areas gradually increases. ×1300.

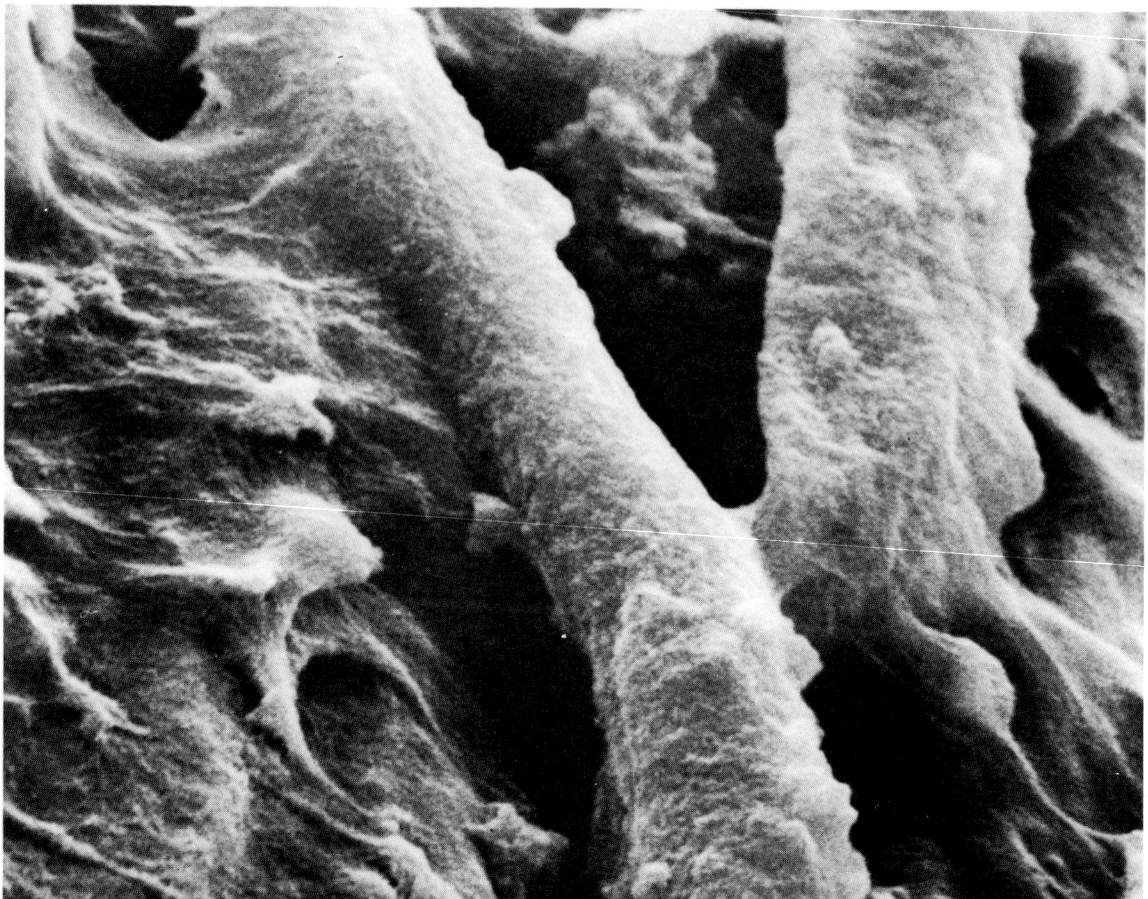

Fig. 3.26 SEM micrograph showing stylar canal with well developing pollen tubes. The accumulated exudate is seen on the surface of each canal cell. ×7000.

3.10 Self-incompatibility (regulation of fertilization)

In many flowers such as lily (Fig. 3.27), brassica, apple and peach, whose pistils and stamens mature together (Fig. 3.28), there is an elegant regulatory mechanism called self-incompatibility for avoiding self-fertilization. In these flowers, even if self-pollination happens to occur, pollen germination on its own pistil is inhibited by the regulation of stigma reaction. The pistils of these flowers refuse entry of the germinating tube or induce the collapse of genetically identical pollen (Fig. 3.29).

Self-incompatibility has been studied by many researchers. However, little is known regarding the mechanism.

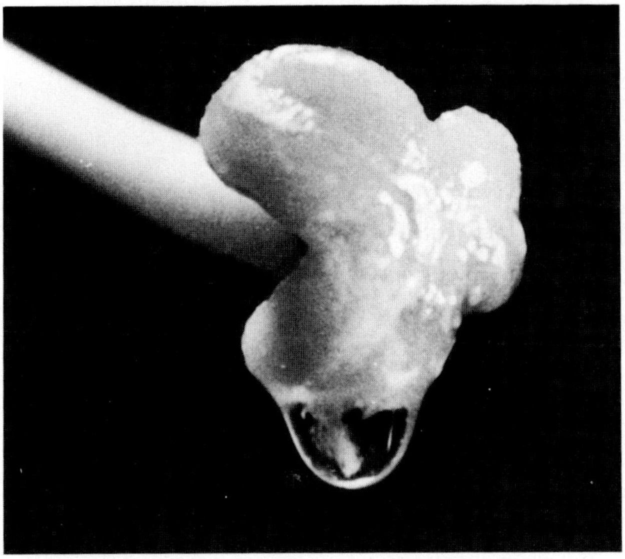

Fig. 3.27 Stigma of lily (*Lilium longiflorum* (secreting mucilage) Photographed by Dr. Hisako Miki-Hiroshige).

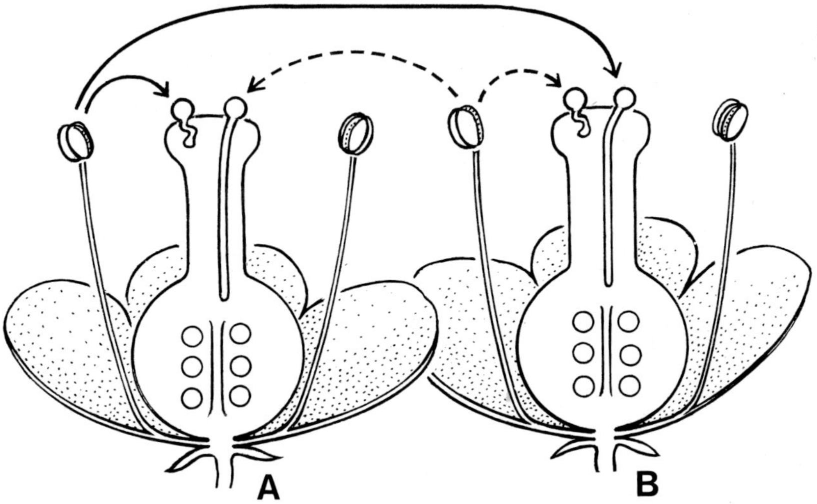

Fig. 3.28 Systems of self-incompatibility. Pollen can germinate on the stigma of other flowers, but not on its own or a relative's.

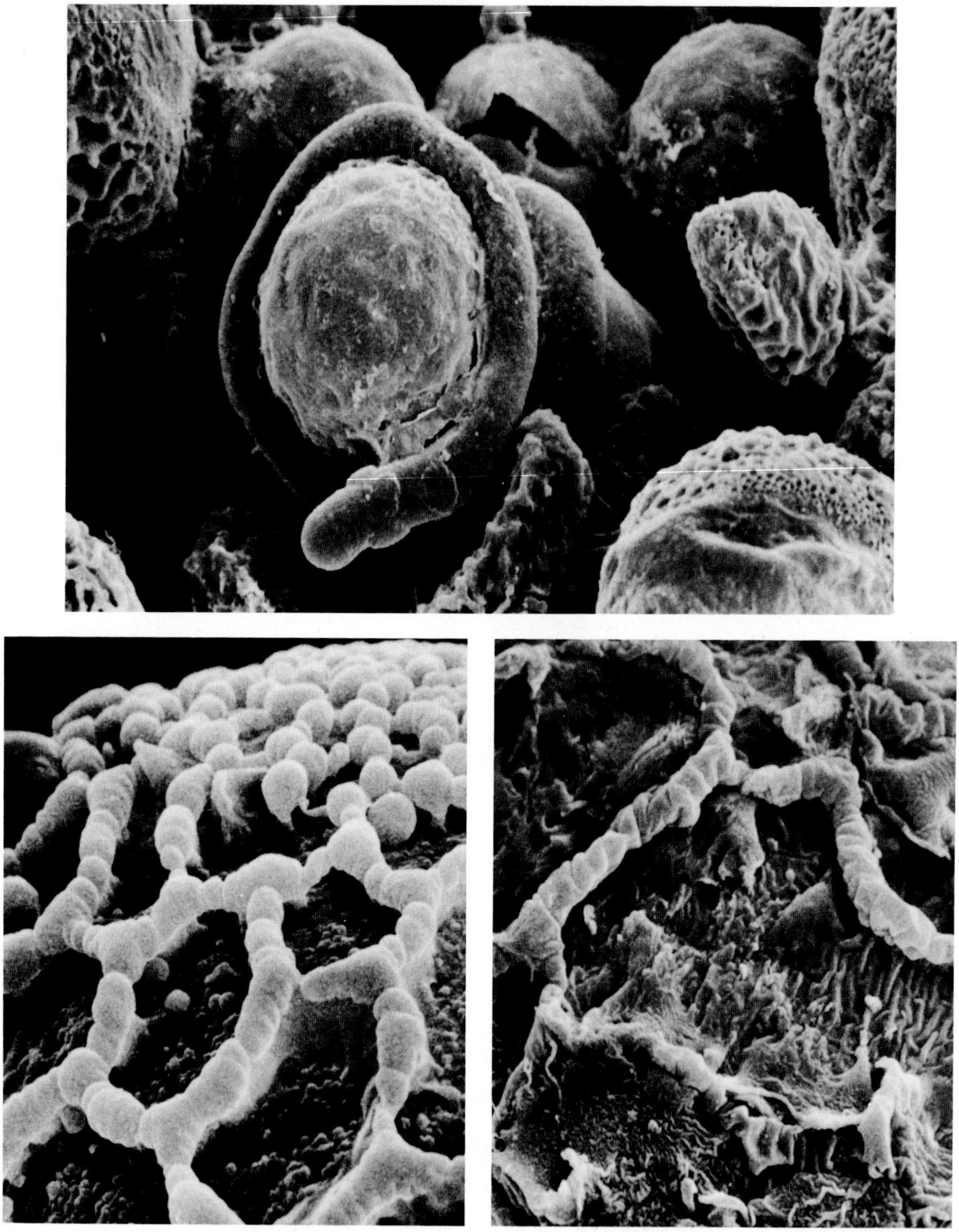

Fig. 3.29 Top: SEM micrograph showing self-incompatible pollen grain of Easter lily with tubes tightly coiled around papillae cells. ×650.
Bottom: SEM micrograph showing compatible pollen grain of Formosa lily on papillate stigma of Easter lily (*left*). ×3000. Germination has proceeded in a normal manner. Surface of the exine shows a normal stratification. SEM micrograph showing self-incompatible pollen grain of Easter lily placed on an anthesis stigma of same flower (*right*). ×3700. Note dissolution of the muri and the wrinkly nature of the enclosure.

4. Physiology of Pollen

Pollen is generated in the anthers of the flower. However, its life is independent from the parent plant body after flowering. It respirates, germinates, grows and dies. Pollen has unique physiological activities differentiated from those of other cells.

4.1 Characteristics of pollen physiology

A pollen grain consists of nuclei, cytoplasm, cell membrane and cell wall. Like other cells, the cytoplasm of pollen contains cytoplasmic organelles, Golgi apparatus, mitochondria and endoplasmic reticulum. As described earlier, a pollen grain consists of two cells, one of which is enclosed within the other larger cell (generative cell and vegetative cell, respectively) in the dinuclear type, and is coated by a hard wall called the exine outside the cell wall. Moreover, unlike other plant cells, pollen cells do not contain chloroplasts, and are consequently not able to conduct photosynthesis.

Therefore, pollen cells, like animal cells, grow heterotrophically obtaining energy sources exogenously by absorbing sugars and amino acids from stigma (Fig 4.1).

Another distinctive characteristic of pollen is that it initiates germination within a short period of time (2–40 minutes) and grows by elongating a pollen tube which is hundreds of times larger than its original size.

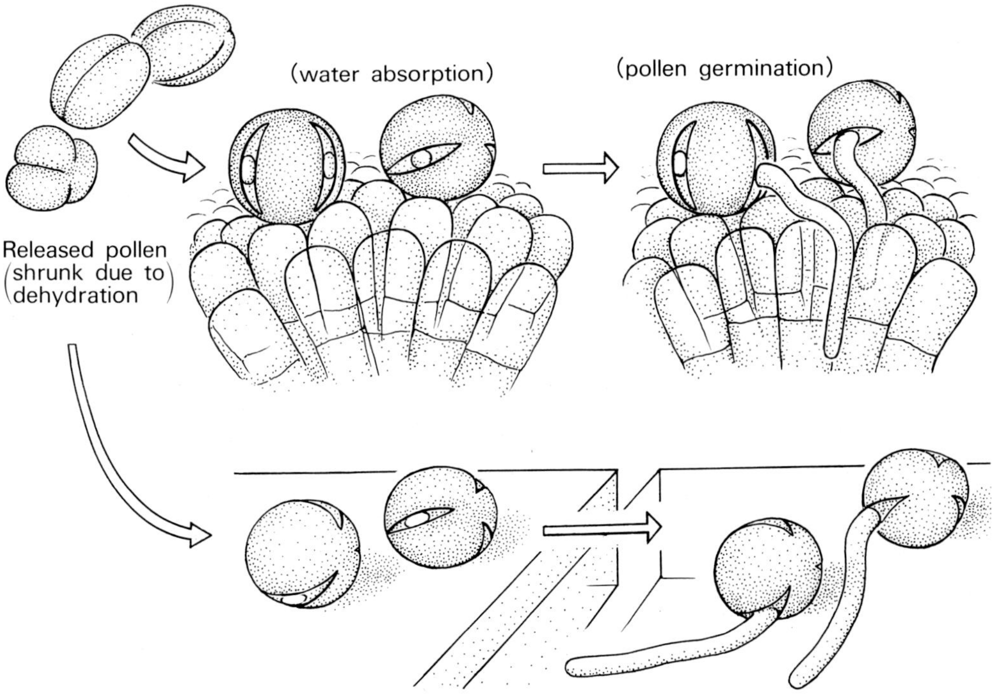

Fig. 4.1 Pollen germination on stigma and culture medium.

4.2 Water absorption and pollen germination

Pollen grains generally shrink due to dehydration after the anthesis stage, and once grains are placed on stigma or artificial medium, they swell due to water absorption. At the beginning of water absorption, the pollen cell membrane is not yet biologically functional, so cell components including sugars, amino acids, and enzymes are excreted from the cell, and, at the same time, large molecular substances enter from outside pollen cell. Gradually, the membrane recovers its biological functions and becomes semipermeable, allowing water to pass through the membrane but preventing substances other than water from entering. Consequently, turgor pressure inside the pollen increases, enabling the tube of the pollen cell to elongate (Figs. 4.2, 4.3, 4.4).

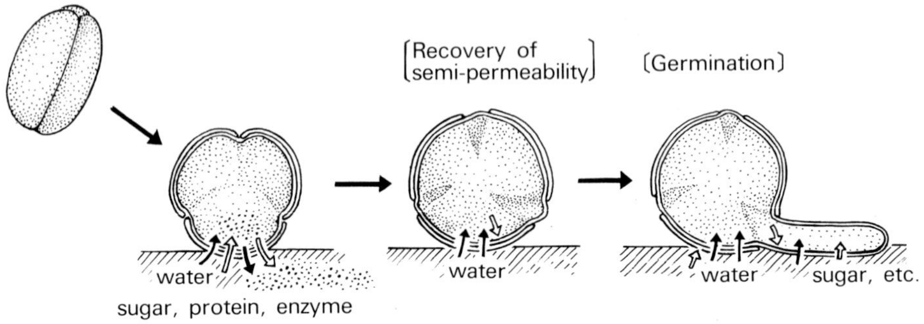

Fig. 4.2 Hydration and transfer of organic molecules through cell membrane for pollen germination.

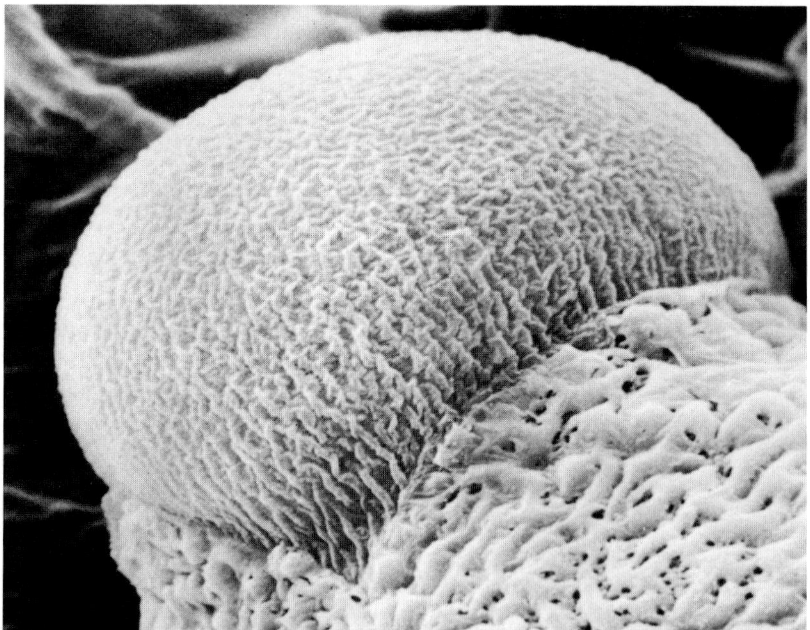

Fig. 4.3 SEM micrograph of germinating pollen grain of *Petunia* showing pollen tube extruded from the aperture. ×4000.

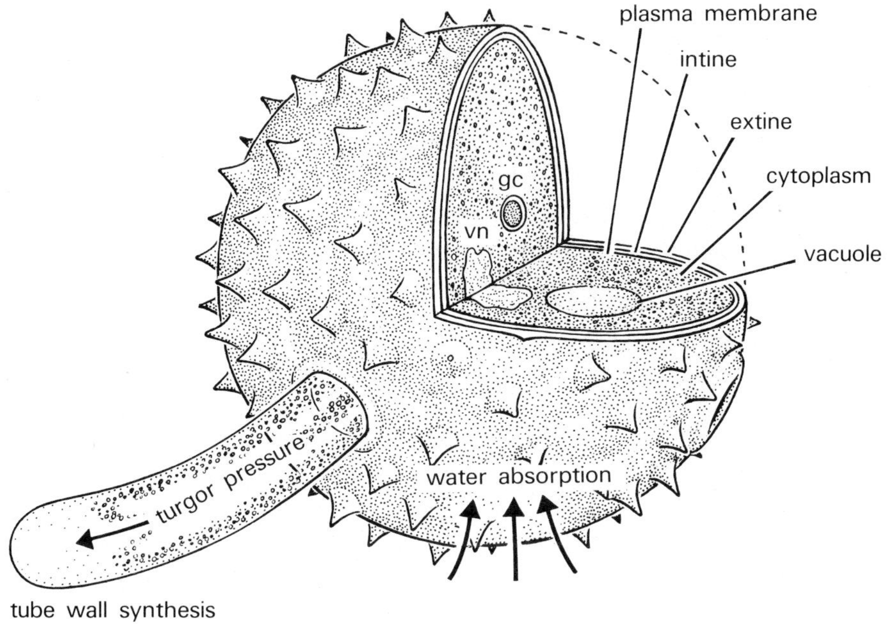

Fig. 4.4 Structure of germinating pollen. Vacuoles are formed in the grain following the translocation of cytoplasm into elongating pollen tube. vn: vegetative nucleus. gc: generative cell.

When the turgor pressure by water absorption and synthesis of tube wall materials are balanced, the pollen cell continues to elongate normally, but if the turgor pressure is too large, the tube breaks and releases plasma; if there is too much synthesis activity, the pollen tube ceases to elongate due to the formation of a thick wall at the tip of the tube.

Duration between the termination of water absorption and tube germination is generally 20–40 minutes but varies among plant species, e.g. two minutes for touch-me-not as the fastest example, and 20 hours for red pine.

4.3 Pattern of pollen tube elongation

The pollen tube elongates as a result of tube wall expansion at the tip upon supplementation of tube wall materials. Thus elongation of the pollen tube occurs only at the tip of the tube, in contrast to the manner of elongation of, for example, elastic bands and seedling root (Fig. 4.5).

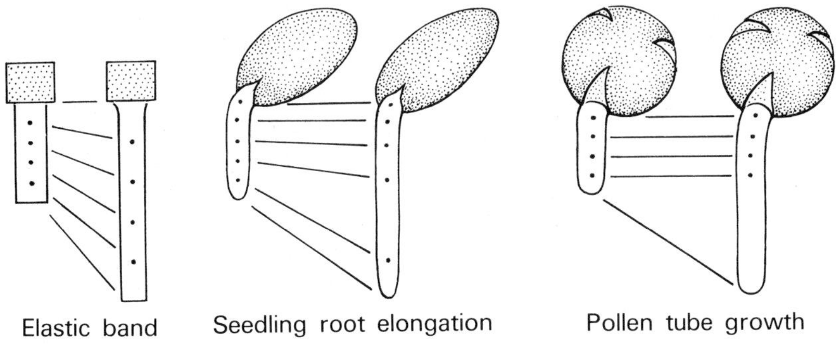

Fig. 4.5 Comparison of elongation patterns among elastic tape, seedling root, and pollen tube.

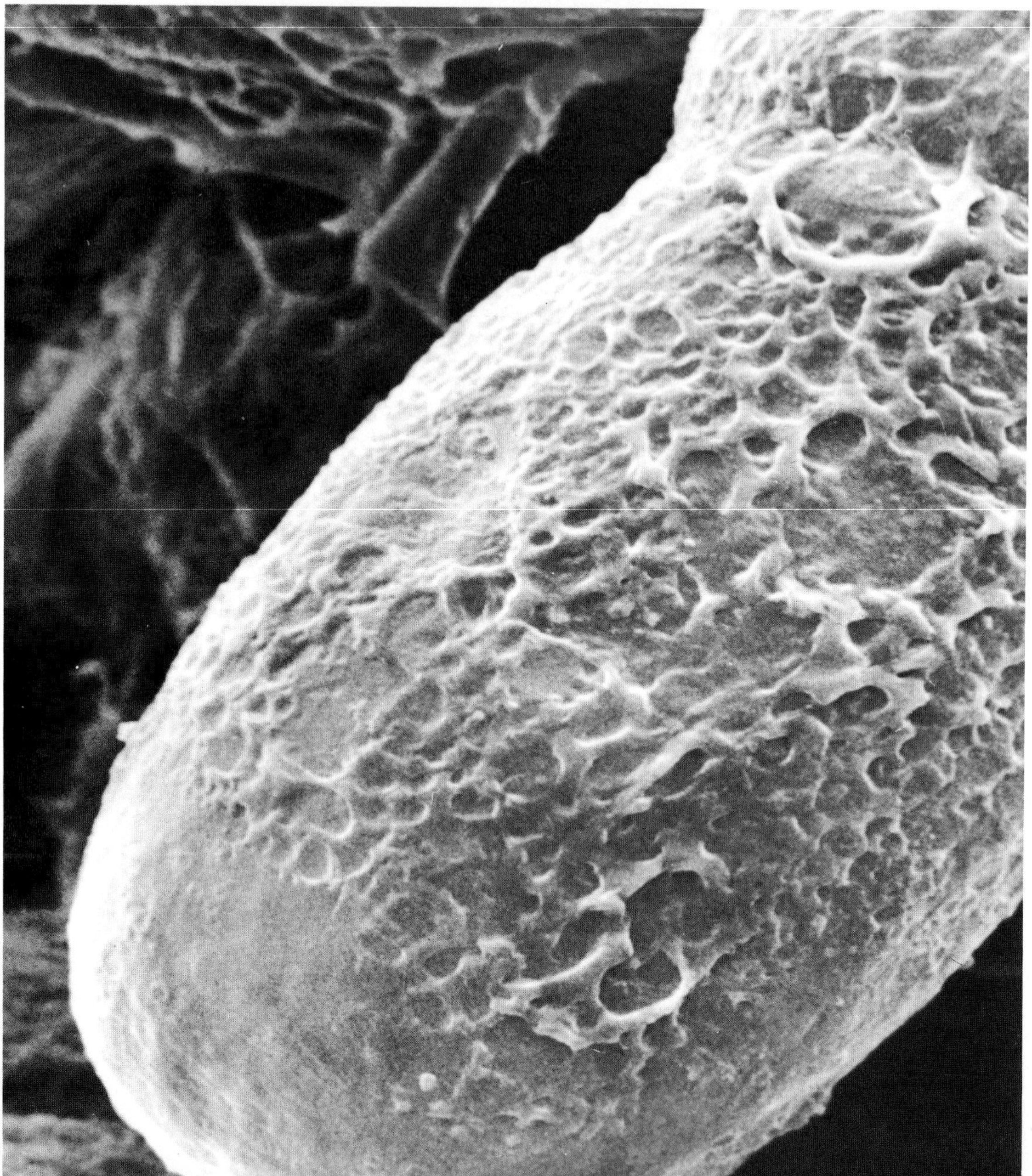

Fig. 4.6 SEM micrograph of pollinated stylar canal of Formosa lily showing the developing tube tip and the canal cell. ×2000.

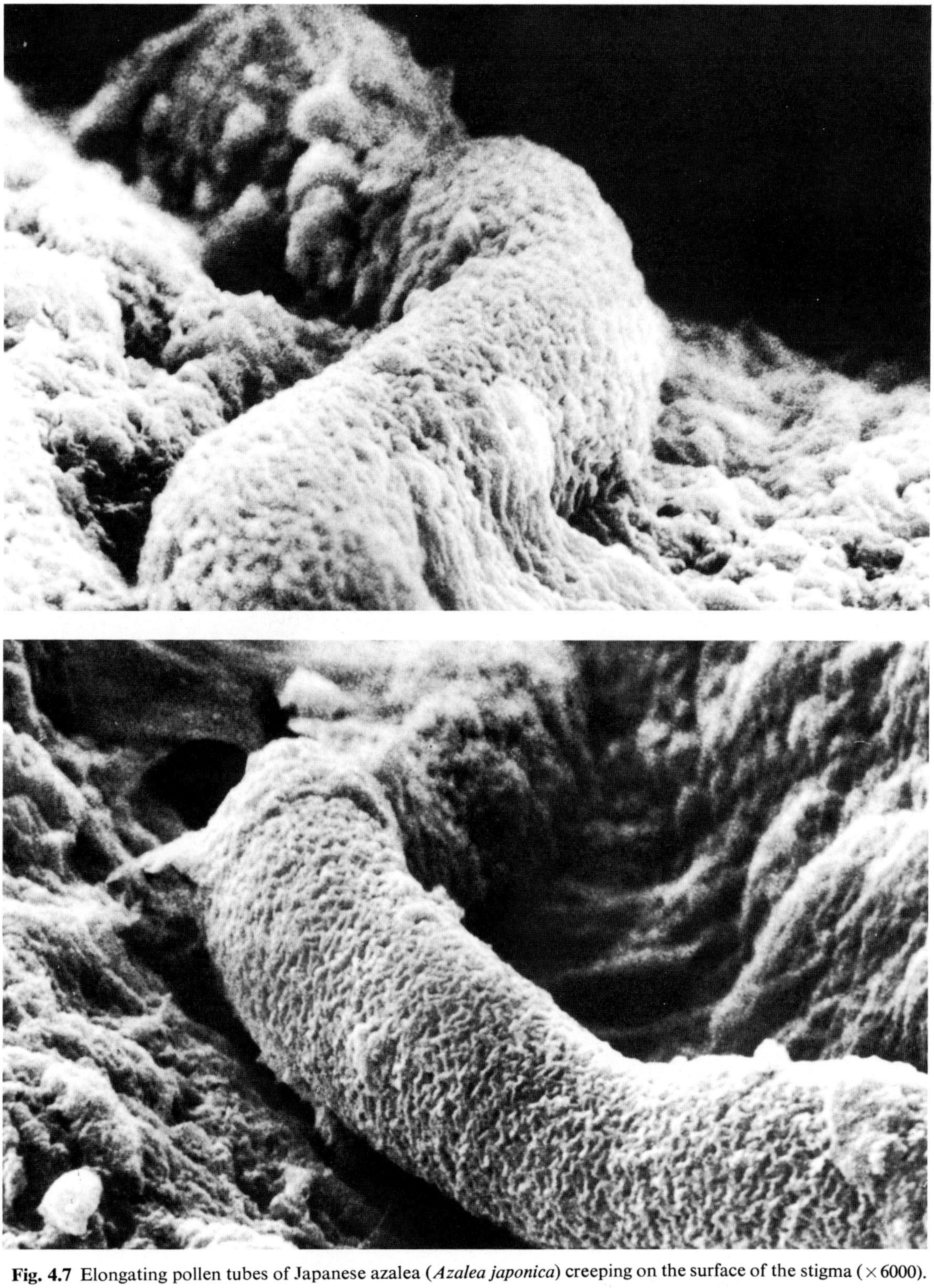

Fig. 4.7 Elongating pollen tubes of Japanese azalea (*Azalea japonica*) creeping on the surface of the stigma (×6000).

Fig. 4.8 Elongating pollen tubes of mountain lily (*Lilium auratum*) (Light micrograph).

To analyze the pattern of tube elongation, we can visually trace the movement of charcoal grains on the tube surface at the tip after spreading charcoal powder on elongating pollen tube.

4.4 Fine structure inside the tube

Under the light microscope, the transparent area identified at the tip of the pollen tube is called the cap block (Fig. 4.9).

Electron micrographs reveal that the cap block contains many Golgi vesicles which have moved to the tip from Golgi apparatus. The tube contains endoplasmic reticulum (ER), mitochondria, Golgi apparatus, starch granules and lipid particles (Figs. 4.9, 4.10).

When germinating pollen is treated with myrmicacim, a biological inhibitor, tube growth is inhibited. The treated tube shows no accumulation of Golgi vesicles in the cap block area. With removal of the inhibitor, vesicles reappear and the tube recovers to elongate (Fig. 4.11).

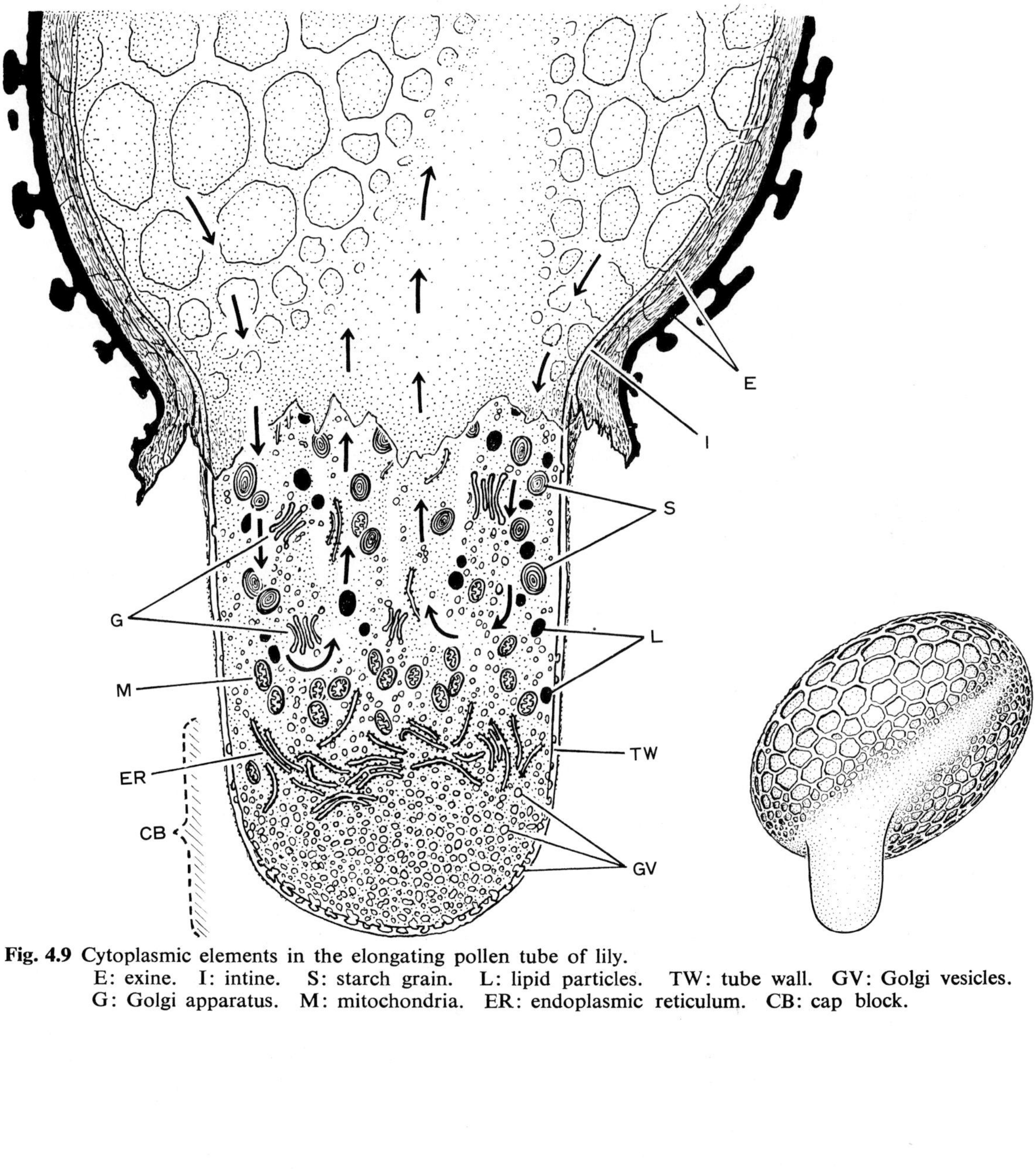

Fig. 4.9 Cytoplasmic elements in the elongating pollen tube of lily.
E: exine. I: intine. S: starch grain. L: lipid particles. TW: tube wall. GV: Golgi vesicles. G: Golgi apparatus. M: mitochondria. ER: endoplasmic reticulum. CB: cap block.

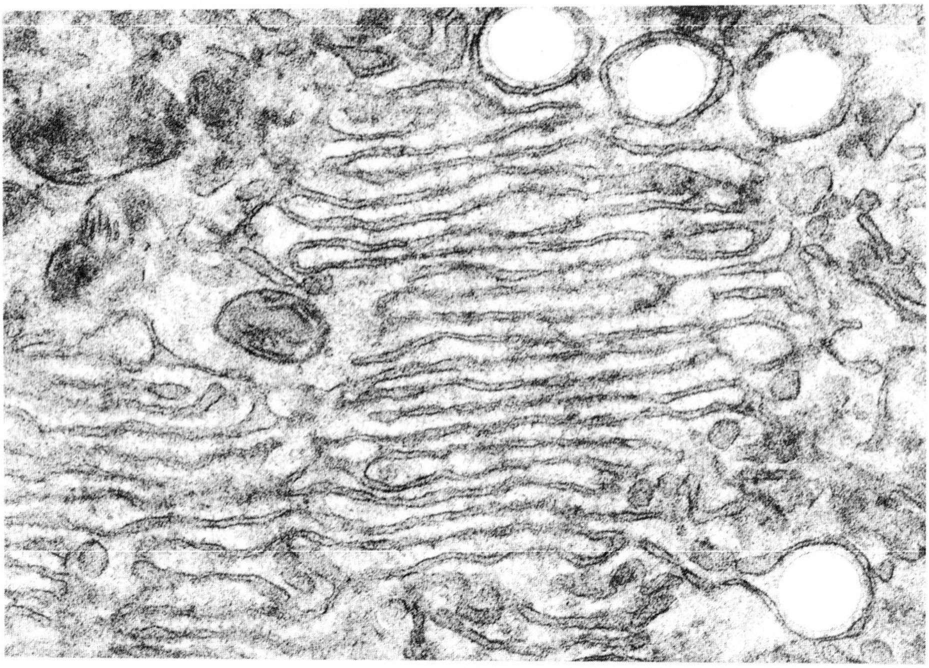

Fig. 4.10 Endoplasmic reticulum (ER: layer structure) and mitochondria (circular structure in upper left) seen in cytoplasm of camellia pollen (CEM).

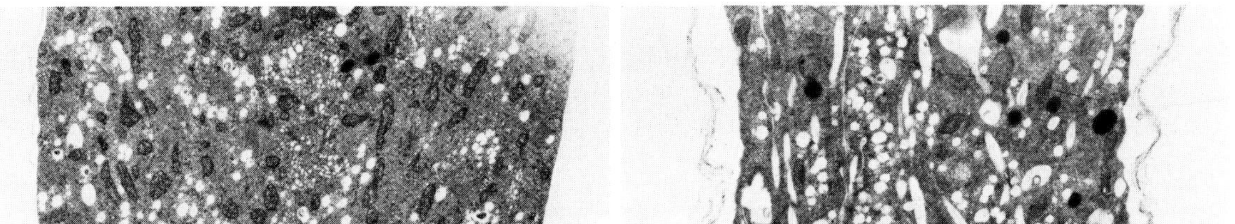

Fig. 4.11 Longitudinal sections of elongating pollen tips of camellia (*Camellia japonica*) (CEM). *Left:* normal growth. *Right:* treated with myrmicacim, an inhibitor.

4.5 Protoplasmic streaming

Protoplasmic streaming occurs in the germinating pollen. Although the precise mechanism has not yet been clarified, apparently the streaming in the growing tube functions to move tube wall materials produced in the Golgi apparatus to the tip. Most of the protoplasm is located in the terminal area of the tube in germinating pollen, and when the tube elongates further, large vacuoles are formed in the grain or basal area of the tube (Fig. 4.12). Around the vacuoles, plasma movement is active through branched paths (Fig. 4.13). When plasmolysis is induced or when callose plug (described below) forms in the pollen tube, the protoplasm separates into one or more blocks in the pollen cell, but the streaming is independently active in each unit (Fig. 4.14). These facts indicate that, unlike the blood stream in animals, protoplasmic streaming in the pollen cell is self-regulating.

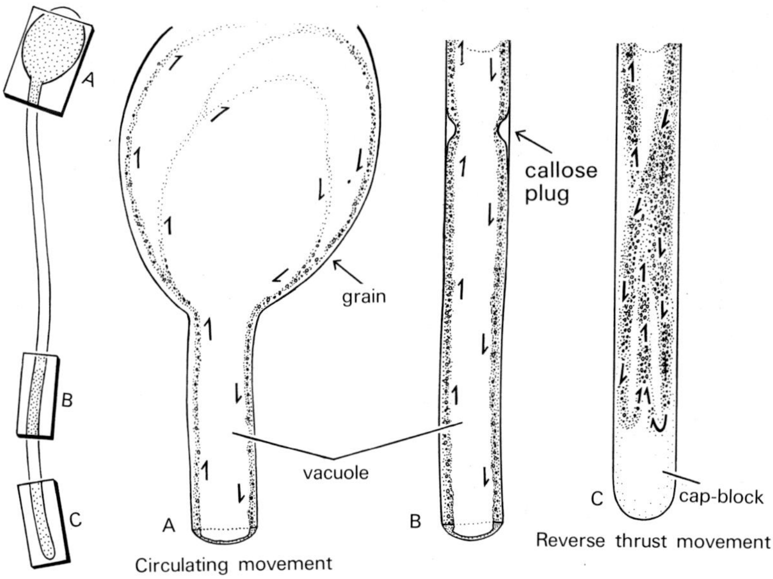

Fig. 4.12 Protoplasmic streaming in germinating pollen.

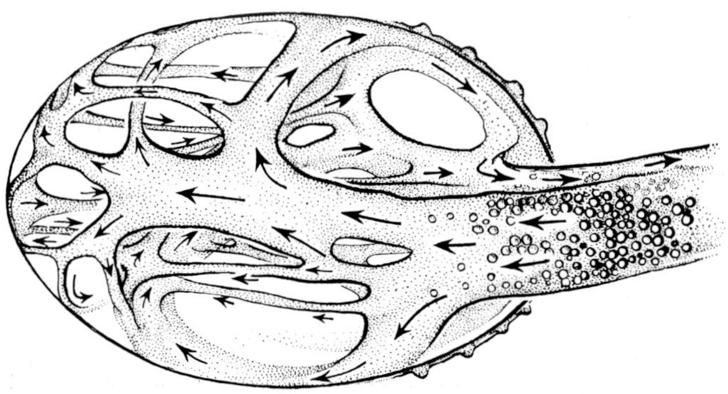

Fig. 4.13 Protoplasmic streaming (circulation) in fully germinated pollen grain.

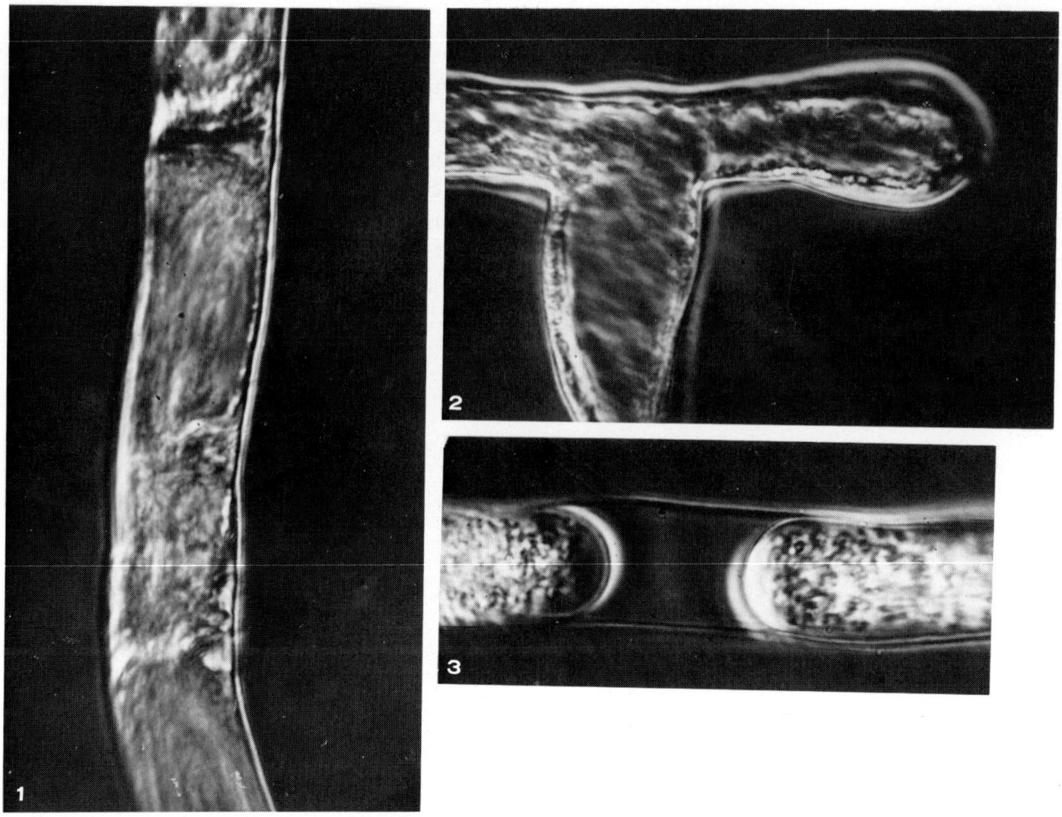

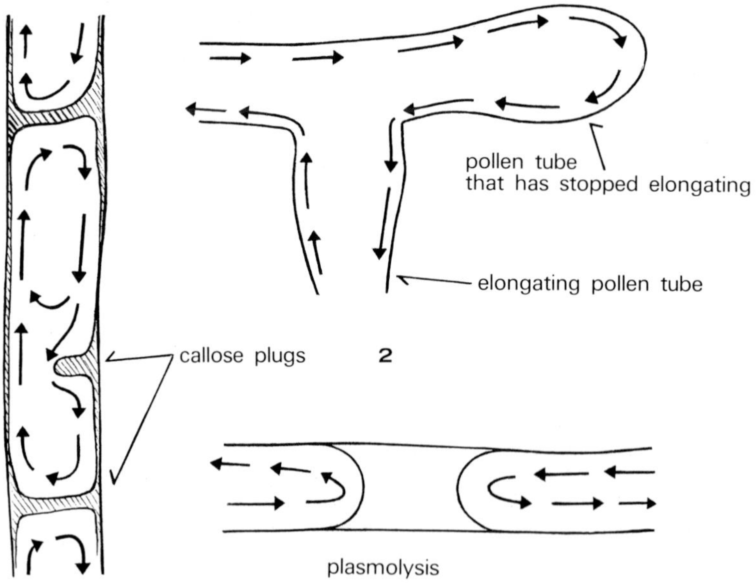

Fig. 4.14 Protoplasmic streaming in elongating pollen tube.
1: pollen tube separated by callose plugs. 2. branching pollen tube. 3. plasmolysis in pollen tube (Cf. illustration below).

4.6 Pollen culture

Pollen can grow not only on the stigma of pistil, but also on culture medium. For example, pollen of touch-me-not and camellia germinate well on artifical 1% agar medium supplemented with 10% sucrose, and elongate long pollen tubes along the surface of the medium (Figs. 4.15, 4.16, 4.17).

When pollen are cultured on medium, various patterns of tube growth can be observed, depending on the culture conditions. In most cases, pollen grow more vigorously when they are sowed thickly rather than thinly (positive density effect). But there are some other species which show negative density effect in artificial pollen culturing (Fig. 4.15).

When scattered on medium, pollen tubes elongate in random directions with bending. On the other hand, when they are arranged in a line, tubes elongate at right angles to the line (Fig. 4.16).

Therefore, it is necessary to chose the proper method for culturing pollen; the random scatter-

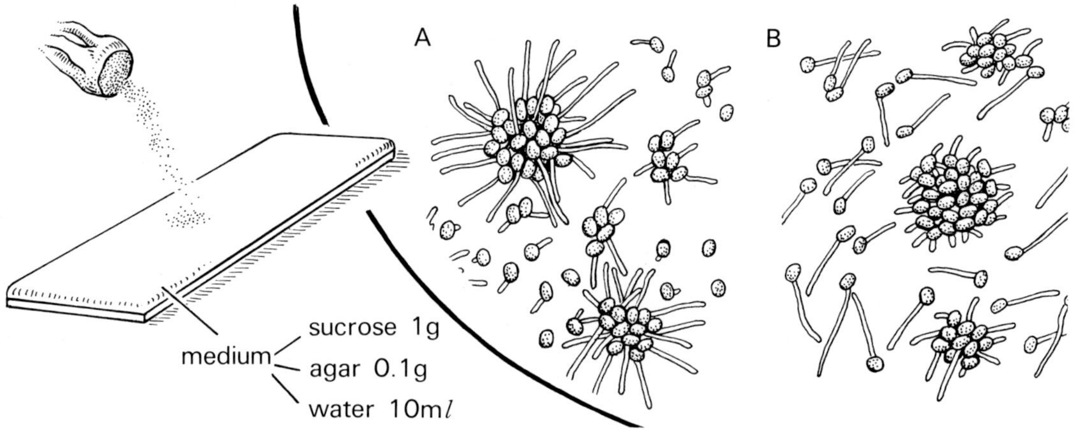

Fig. 4.15 Pollen culture (*left*), and density effects of polen germination (*right*). A: positive density effect. B: negative density effect.

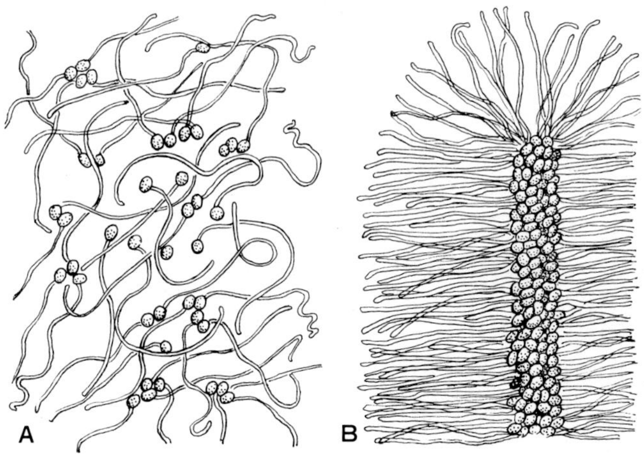

Fig. 4.16 Methods of pollen culture. A: scattering methods for test of germination frequency. B: line-up method for test of tube elongation.

ing method for checking germination frequency and the line-up method for measuring the length of pollen tube.

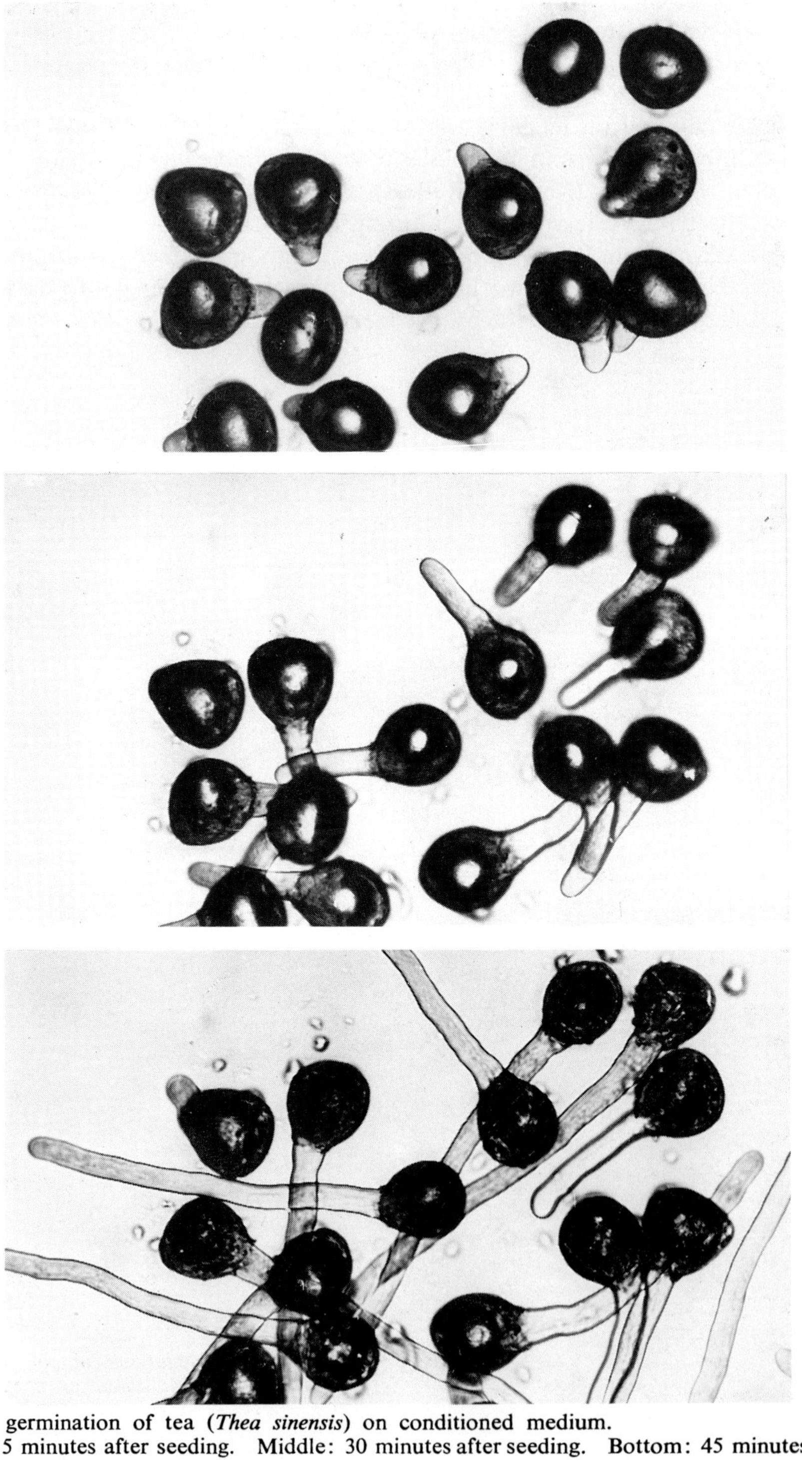

Fig. 4.17 Pollen germination of tea (*Thea sinensis*) on conditioned medium.
Top: 15 minutes after seeding. Middle: 30 minutes after seeding. Bottom: 45 minutes after seeding.

4.7 Movement of nuclei and formation of callose plug

The nuclei of pollen by themselves do not migrate in the pollen cells or tubes, but move with the cytoplasm toward the tip of the pollen tube (Fig. 4.19). Nuclear division of the generative nucleus takes place during movement in the pollen tube to produce two sperm cells in the case of dinuclear pollen (Figs. 4.18, 4.19).

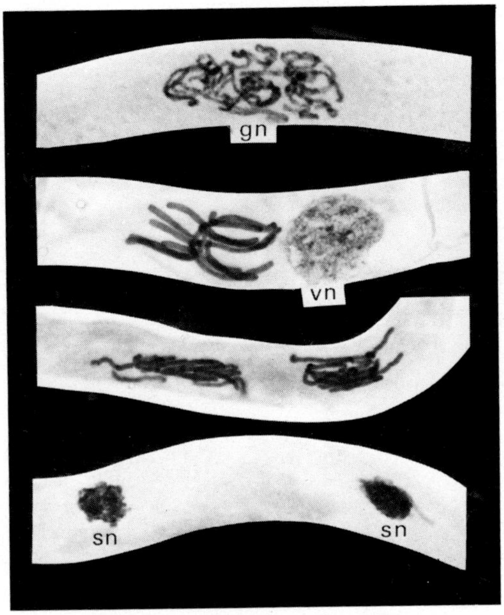

Fig. 4.18 Pollen mitosis of hyacinth (*Hyacinthus orientalis*). (photographed by Mr. A. Kamijo). gn: generative nucleus. vn: vegetative nucleus. sn: sperm nucleus.

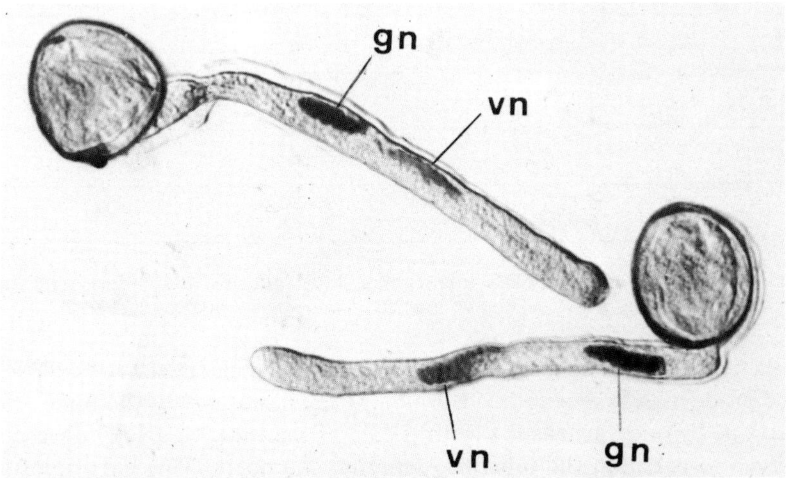

Fig. 4.19 Vegetative nucleus (vn) and generative nucleus (gn) in elongating pollen tube of hyacinth (*Hyacinthus orientalis*).

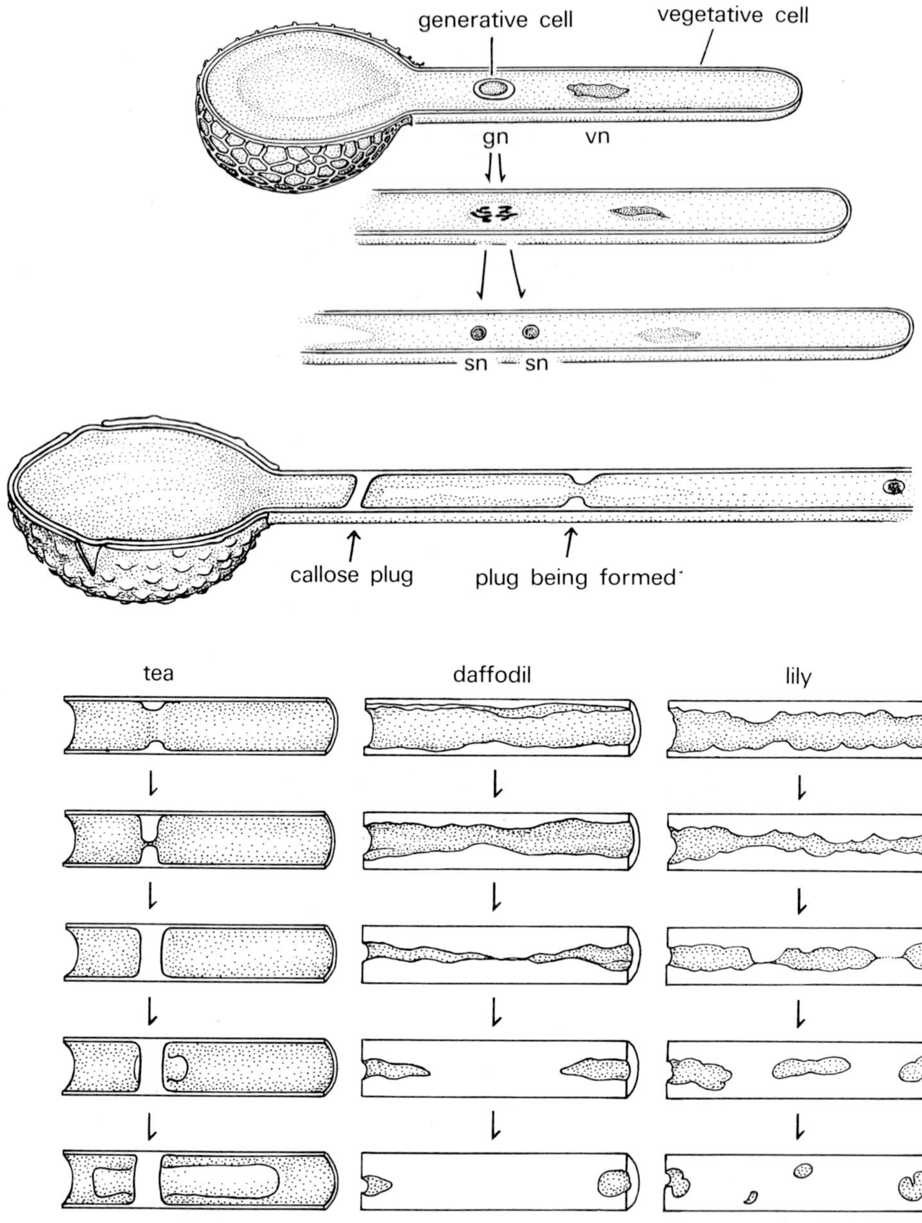

Fig. 4.20 Division of generative cell in pollen tube (pollen tube mitosis), and callose plug formation. gn: generative nucleus. vn: vegetative nucleus. sn: sperm nucleus

When the pollen tube elongates to a certain extent, plugs which consists of callose, a polysaccharide, are formed inside it. A first plug is generated at the basal part of the tube, and several plugs are then formed one by one towards the central part of the tube. In a fully developed tube, many plugs are observed to partition the tube into separate chambers. This partition functions to prevent the backflow of cytoplasm and nuclei in the long pollen tube (Fig. 4.20). This mechanism reminds us of the ship transport system of the Panama canal.

4.8 Promotion and inhibition of pollen tube elongation

Promotion and inhibition of tube elongation are examined by observing germination frequency or by measuring tube length, where pollen grains are sown on supplemented medium. Compared with the control, the inhibitory or promotive effects of a certain substance can be evaluated in this way (Fig. 4.21). Boron (or boric acid) and Ca^+ ($CaNO_2$) generally promote pollen growth; vitamins B_1 and B_6 and some amino acids are also effective for promotion.

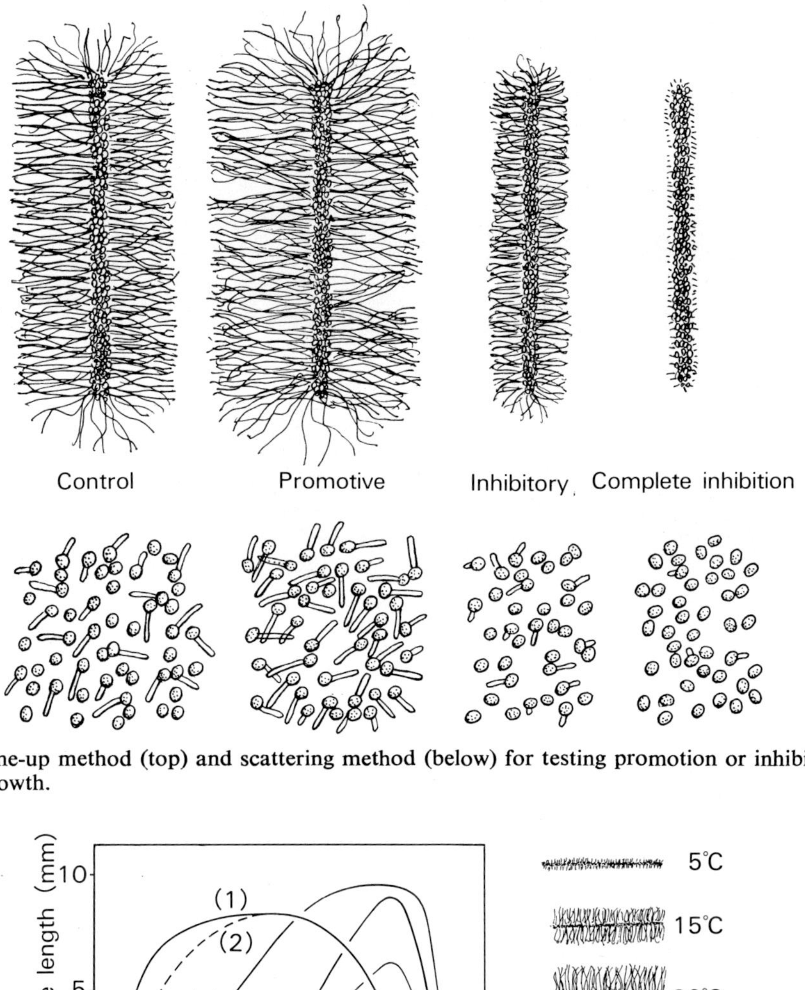

Fig. 4.21 Line-up method (top) and scattering method (below) for testing promotion or inhibition of pollen tube growth.

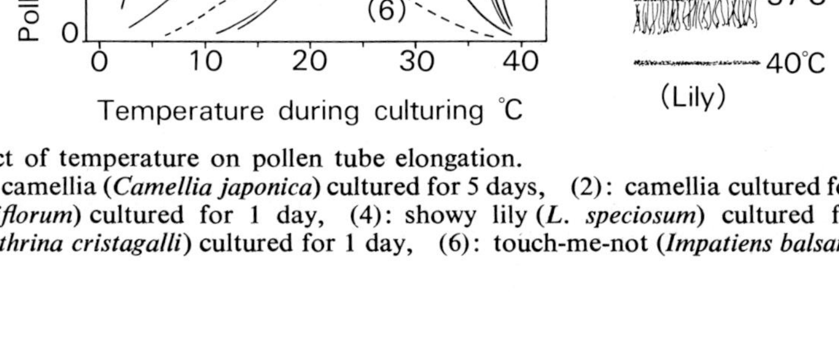

Fig. 4.22 Effect of temperature on pollen tube elongation.
(1): camellia (*Camellia japonica*) cultured for 5 days, (2): camellia cultured for 2 days, (3): lily (*Lilium longiflorum*) cultured for 1 day, (4): showy lily (*L. speciosum*) cultured for 1 day, (5): coral tree (*Erythrina cristagalli*) cultured for 1 day, (6): touch-me-not (*Impatiens balsamina*) cultured for 1 day.

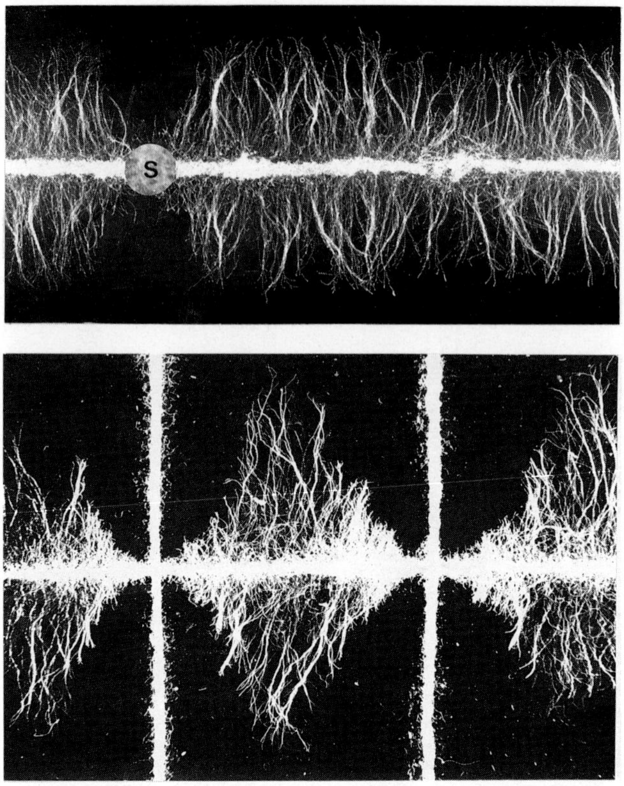

Fig. 4.23 Inhibitory effects of stigma and pollen of other species on pollen growth.
Top: Inhibition of germination of camellia pollen around the excised dish of lily stigma (S).
Bottom: Inhibition of tube elongation of mountain lily pollen (horizontal line-up) near the array of sasanqua pollen (longitudinal line-up).

The optimal temperature condition is 29–30° C for pollen growth, while generally under conditions of over 35° C or less than 10° C, germination and tube elongation are inhibited or suppressed. However, pollen of camellia, which flowers in winter, can germinate at even below 5° C (Fig. 4.22).

If excised pieces of pistil are placed near pollen grains on medium, promotion or inhibition of pollen development is observed depending on the species. This is due to secretion of the controlling substances from the pistil. In general, promotion of pollen growth is observed in the pollen-pistil combination between among related species, while inhibition is observed between unrelated species (Fig. 4.23). Pollen interaction can occur between different species. When two kinds of pollen grains are mixed together on a medium, germination or tube elongation is inhibited in certain combinations. For example, Fig. 4.23 (bottom) shows pollen growth of mountain lily depressed by pollen of sasanqua tea when both are arranged on the medium at right angles to the lines.

4.9 Effects of volatile substances from plant tissues (1)

Pollen growth is influenced by volatile substances in air. Fig. 4.24 shows an experimental system for analyzing the effects of volatile substances on pollen growth from wasabi (Japanese horse radish), garlic, onion and lemon, all of which release aroma into the air.

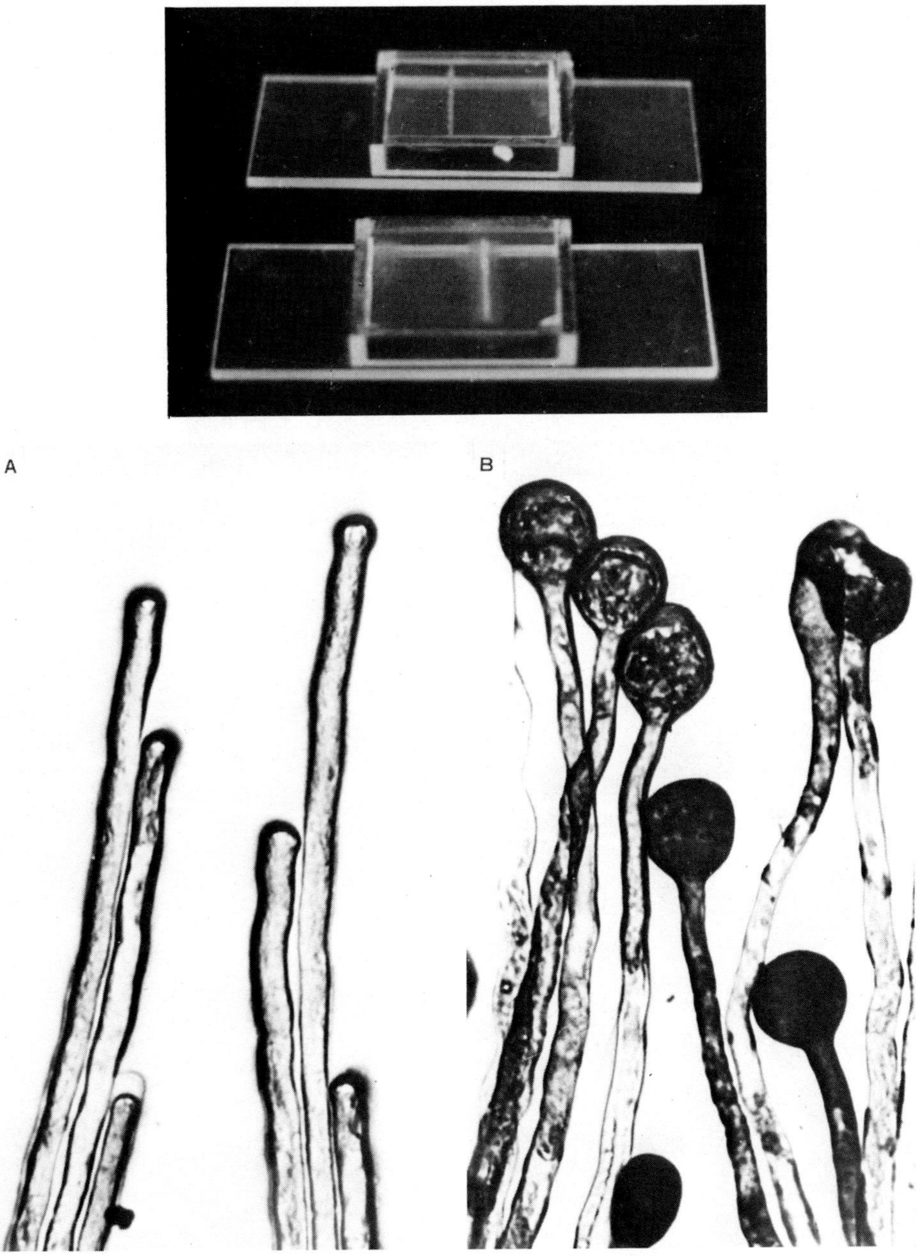

Fig. 4.24 Effect of volatile substances on pollen growth (1).
Top: incubation chamber on glass slide for pollen germination. A piece of plant tissue is placed on the chamber (white material in photo.).
A: elongating tubes of camellia pollen (control). **B**: elongating pollen tubes treated with volatile substances from onion pieces (for 5 minutes), showing malformation at their tips.

An incubation chamber was built on a slide glass mounted with agar medium on which sliced or grated pieces of the above plant species were placed, and pollen germination was examined.

Malformation of the tip of elongating pollen of camellia, induced by treatment with volatile substances from onion slices for five seconds, is shown in Fig. 4.25. This type of tube malformation

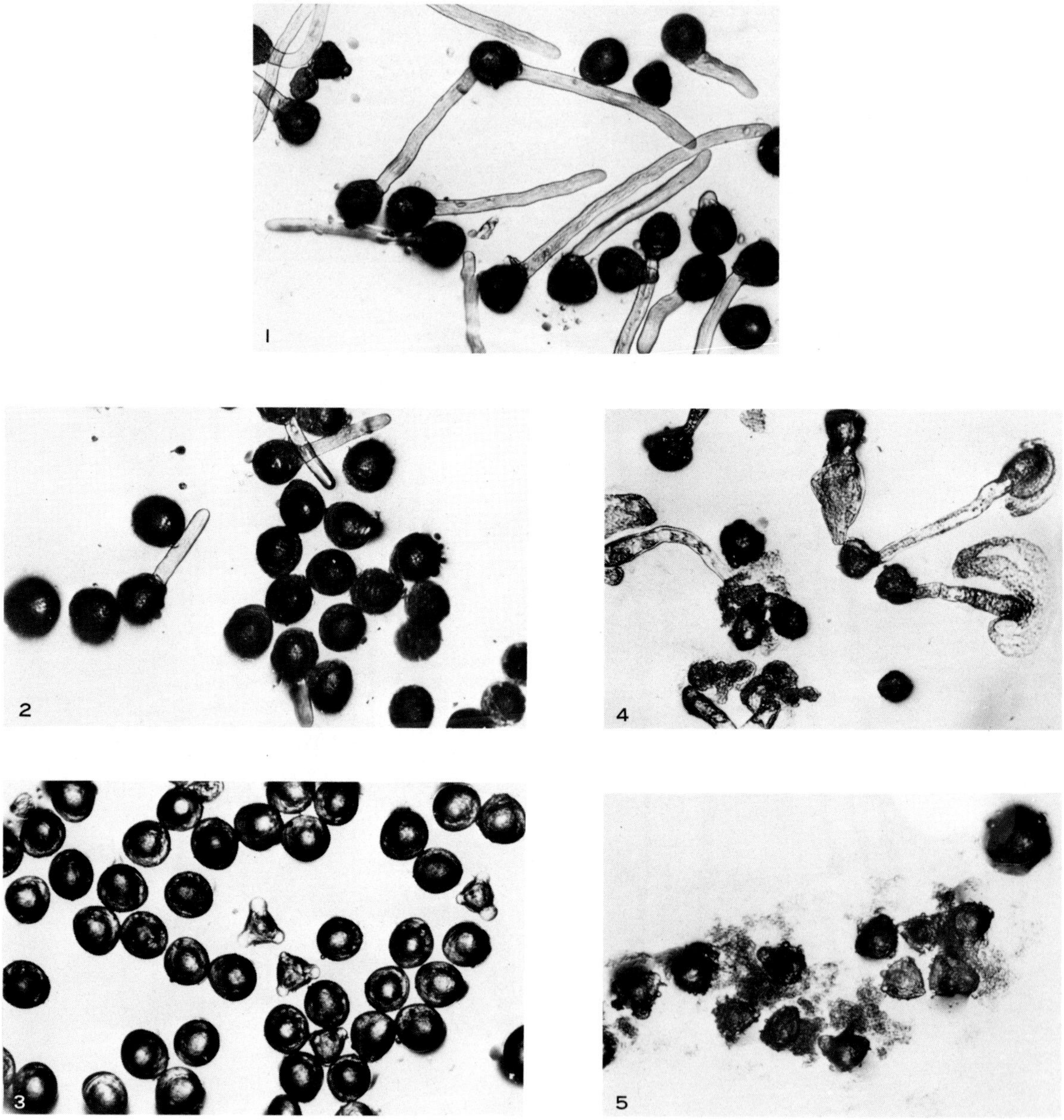

Fig. 4.25 Effects of volatile substances on pollen growth (2).
1: growth of camellia pollen (control)
2: treated with 0.05 g grated Wasabi (Japanese horseradish).
3: treated with 0.1 g grated Wasabi.
4: treated with 0.01 g lemon peel.
5: treated with 0.05 g lemon peel inducing collapse of grains.

is also caused by irradiation with γ-rays of 200 kr on elongating pollen. It is noteworthy that the volatile substances from plant tissues have effects similar to those of γ-ray irradiation on pollen growth.

It should be mentioned that ethylene gas, which inhibits stem elongation of seedling growth at 5–10 ppm, does not affect pollen development; pollen germinate well and pollen tubes elongate under 10% ethylene gas in air.

4.10 Effects of volatile substances from plant tissues (2)

We have developed an experimental system for examining the effects of volatile substances from plant tissues on pollen germination, as illustrated in Fig. 4.26. Certain amounts of chopped (2 mm^3) or grated pieces of various plant tissues are placed in a plastic chamber (1 *l*), in which pollen germination and pollen tube mitosis on agar medium are examined under exposure to volatile substances from the plant tissues. In Fig. 4.27, some results of these experiments are summarized; length of pollen tubes of camellia (*Camellia japonica*) was measured after incubation for 20 hours under exposure to volatile substances from 15 kinds of plant species. Without plant tissue, pollen tube elongated by 7 mm on average under the employed condition, while the grated tissues of rhizome of wasabi (*Eutrema wasabi*), root of radish (*Raphanus sativus*), tuber of garlic (*Allium sativum*), bulb of onion (*Allium cepa*), and leaf of steak plant (*Perilla frulescens*) severely inhibited pollen tube elongation. Peels of lemon (*Citrus limon*) and orange (*Citrus sinensis*), or leaves of leek (*Allium tuberosum*) and dokudami (*Houttuynia cordata*) had moderate inhibition effect, but leaves of pine (*Pinus thunbergii*), tradescantia (*Tradescantia reflexa*) and camellia (*Camellia japonica*) showed no effect on pollen germination. The experiment also showed that the grated pieces of the tissues had greater inhibitory effect than chopped ones, indicating that the smaller pieces volate more substances given their larger surface area.

Eighteen chemical components of lemon peel were analyzed separately for inhibition ability on pollen germination in the same system as that shown in Fig. 4.26, except that 0.5 mg of each chemical was placed in a petridish instead of plant pieces. The experimental results, as shown in Fig. 4.28, indicate that the inhibition is due to the chemical nature of certain aldehydes (which have a

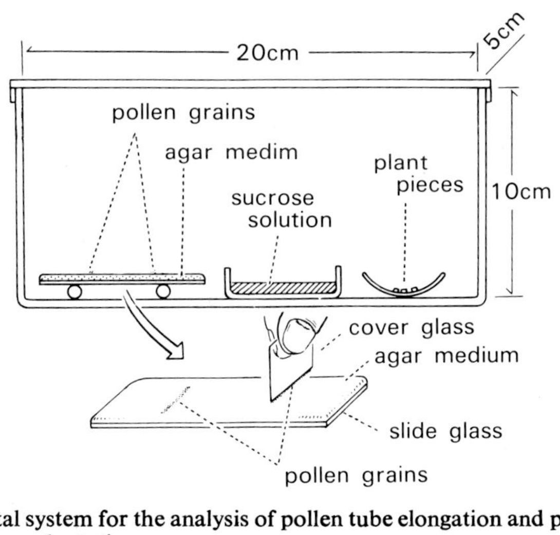

Fig. 4.26 Experimental system for the analysis of pollen tube elongation and pollen mitosis treated with volatile substances from plant tissues.

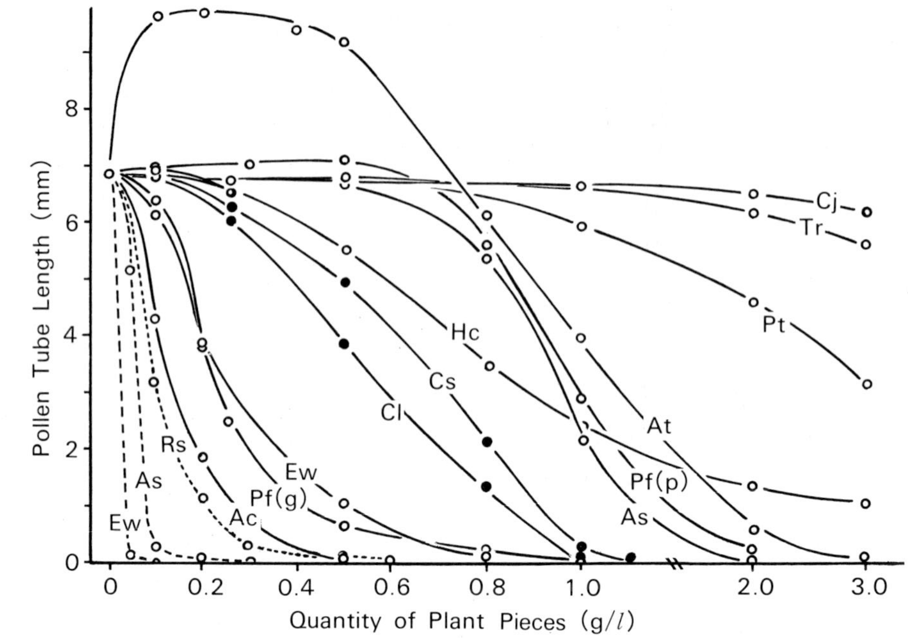

Fig. 4.27 Effect of volatile substances from plant tissues on pollen tube elongation of camellia. cj: camellia (leaf). tr: tradescantia (leaf). pt: pine (leaf). pf(g): green beefsteak plant (leaf). pf(p): purple beefsteak plant (leaf). at: leek (leaf). hc: dokudami (leaf). cs: orange (pericarp). cl: lemon (pericarp). ew: wasabi (rhizome) ac: onion (bulb) rs: radish (root) as: garlic (tuber). ——: chopped pieces. ------: grated tissues.

–CHO base), citral, citronellal, decyl-aldehydes, which are volatile substances included in lemon oil.

These volatile substances also inhibit growth of bacteria and fungus. For example, if a piece of bread is kept together with pieces of wasabi or horseradish, no fungus propagation is detected on the bread. In the future, it is possible that a "flavor stocker" filled with volatile substances from plants to keep food fresh will replace the refrigerator.

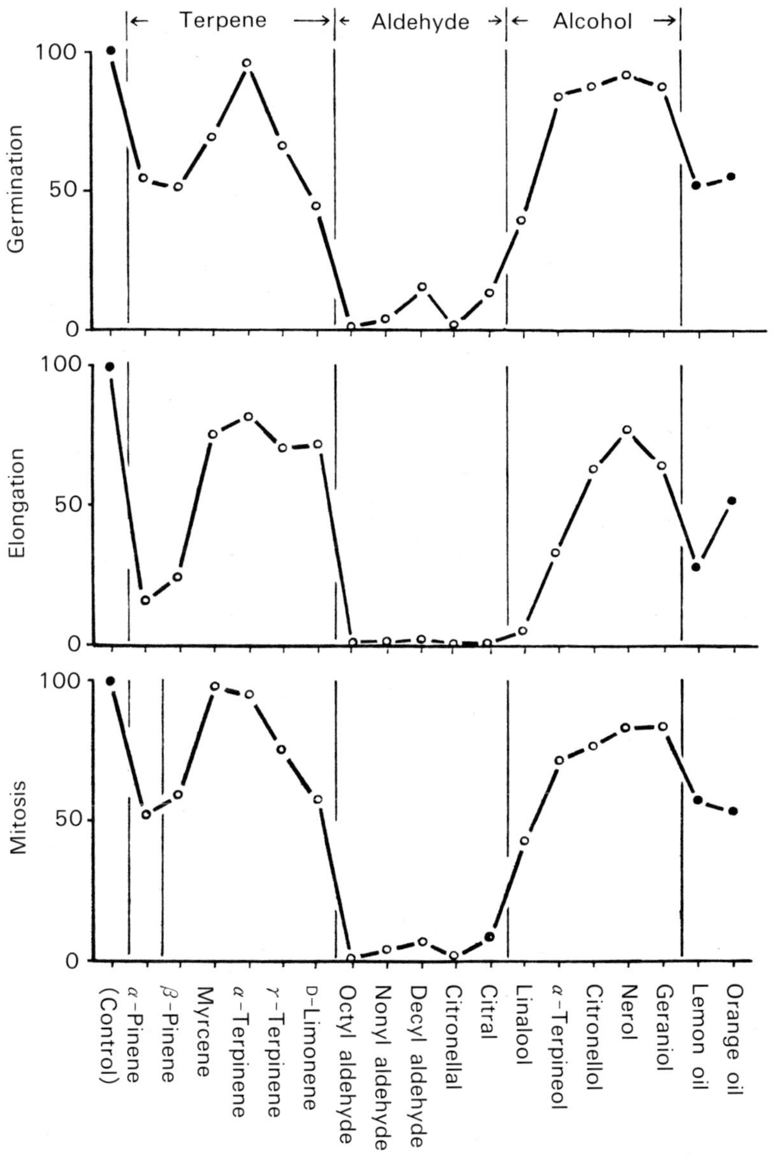

Fig. 4.28 Effects of lemon and orange oils and their components on pollen germination, tube elongation and tube mitosis of camellia.

4.11 Longevity of pollen in organic solvents

It is generally said that organic solvents such as alcohol, ether, chloroform, and phenol are fatal to organisms. Consequently, biologists usually keep their materials away from these solvents. However, it has recently been demonstrated that viability of organisms can be maintained when stored in these organic solvents under certain conditions. We have demonstrated that pollen grains can germinate on medium or even participate in fertilization after they have been kept in these organic solvents (Fig. 4.29). We examined 50 kinds of organic solvents for their storage effect on pollen viability after pollen grains of camellia were kept in them for one week. The results indicate that glacial acetic acid is the only solved completely fatal to pollen germination in camellia. The petunia plants shown in Fig. 4.30 developed from a seed obtained from pollination with pollen grains stored in ether for one week. It does not show any abnormality.

According to Dr. N. Hirata of Chiba University, pollen grains of pear, stored for one year in ether are used for fertilization with flowers in the following spring, and he obtained many fruits which were bigger and sweeter than those obtained by fertilization with pollen stored under dry and cold conditions (Fig. 4.31).

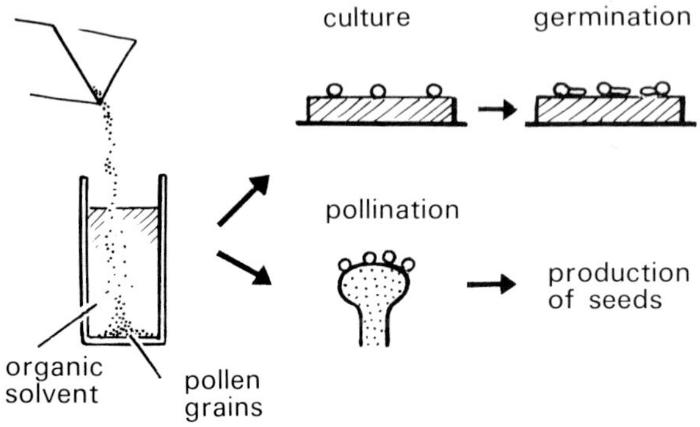

Fig. 4.29 Pollen viability after storage in organic solvents.

Fig. 4.30 Petunia plants obtained from the cross with pollen stored in ether for one week.

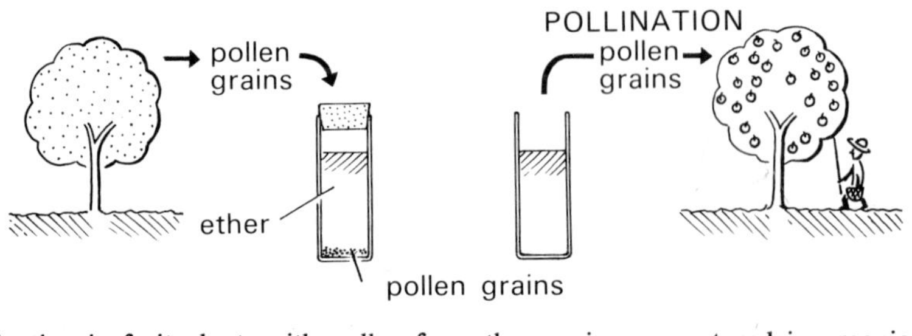

Fig. 4.31 Pollination in fruit plants with pollen from the previous year stored in organic solvent for one year.

Although there remain many problems to be solved in this kind of study, the day will come when pollen storage on a semi-eternal basis is a reality. Fig. 4.32 shows germination of lily pollen after storage for ten years in ethylalcohol. These pollen grains are capable not only of germination but also fertilization (Fig. 4.33).

We have also recently demonstrated that these organic solvents can conserve seed viablity of plants and hatching ability of fertilized eggs of brine shrimp.

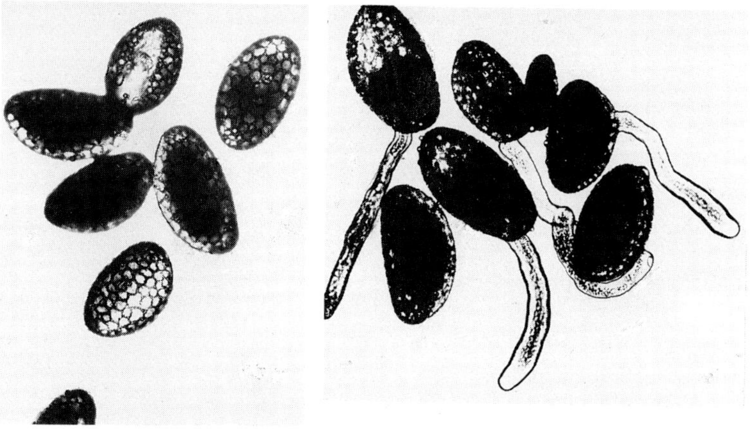

1 2

Fig. 4.32 Germinability of pollen grain of lily stored for ten years under dry and cold conditions (1), and in butyl-alcohol (2).

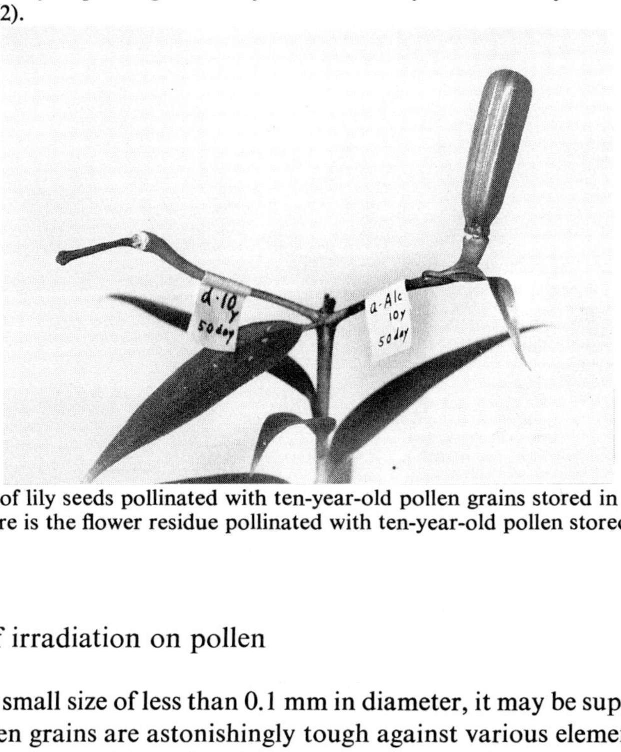

Fig. 4.33 Fruiting of lily seeds pollinated with ten-year-old pollen grains stored in butylalcohol. The left branch in the picture is the flower residue pollinated with ten-year-old pollen stored under dry and cold conditions.

4.12 Effect of irradiation on pollen

Because of its small size of less than 0.1 mm in diameter, it may be supposed that pollen is fragile. The fact is, pollen grains are astonishingly tough against various elements. For example, less than 1 kr of X-ray irradiation is lethal to humans or plants, while pollen is generally tolerant to more than 100 kr of irradiation. Fig. 4.34 presents a summary of irradiation experiment on pollen germinability with varying doses of γ-rays. Lethal doses (LD_{50}: dose lethal to 50% of treated samples) of pollen germination by γ-ray are 480 kr for lily, 420 kr for camellia, and 300 kr for touch-me-not. Considering that 100 kr of γ-rays turns window glass brown, this durability of pollen far surpasses the organic level (Fig. 4.35). Irradiated pollen, however, do not undergo generative nucleus division, and lose fertilization ability.

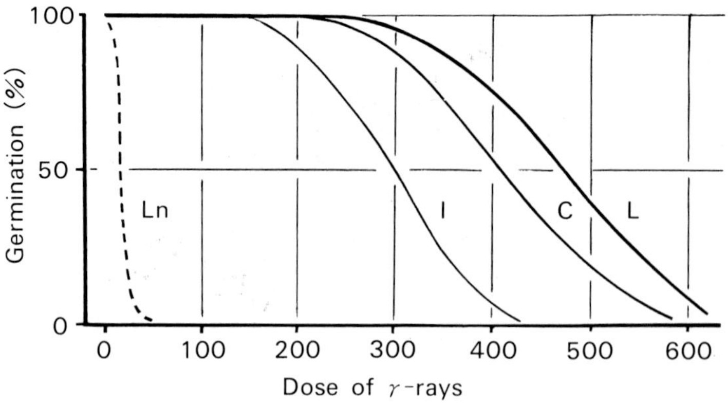

Fig. 4.34 Effects of γ-ray irradiation on pollen.
L: pollen germinability of mountain lily (*Lilium auatum*).
C: pollen germinability of camellia (*Camellia japonica*).
I: pollen germinability of touch-me-not (*Impatiens balsamina*).
Ln: nuclear division of pollen of mountain lily

Fig. 4.35 Pollen germination of camellia after irradiation with 200 kr γ-ray.

4.13 Resistance to heat and decomposition

Pollen is very strong against temperature. Pollen of cyclamen can germinate on media after treatments with 80° C for one hour and 100° C for ten minutes (Fig. 4.36). Heat tolerance of pollen grain may be due to the fact that they have completed initiation of germination and are under a dehydrated condition at the anthesis stage.

As noted above, in spite of their small size of less than 0.1 mm in diameter, pollen grains are amazingly durable. The outer wall (exine) of pollen grain, which consists of sporopollenin, is chemically very stable and is not decomposed by vitric acid, hydrochloric acid, caustic soda, or even by hydrochloric fluorine which can dissolve rocks and glass crystals. Because of this durability, exine of pollen grains have been found intact in strata millions of years old (see Fig. 4.37).

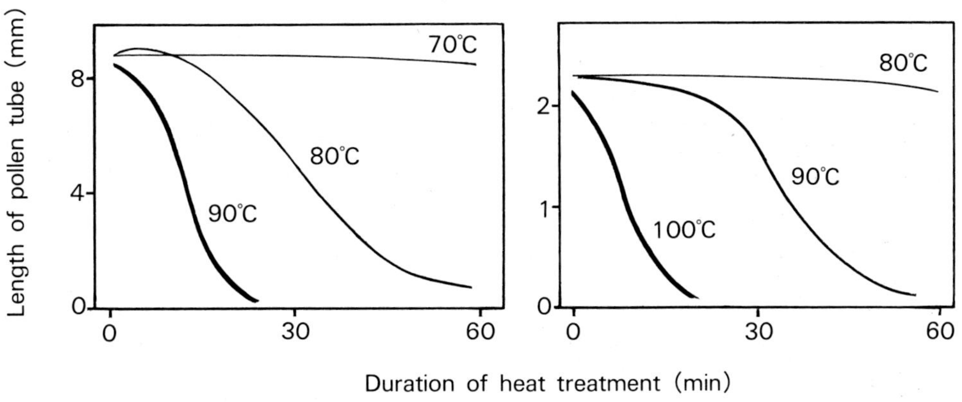

Fig. 4.36 Heat tolerance of pollen.
Left pollen tube elongation of camellia. *Right:* pollen tube elongation of cyclamen

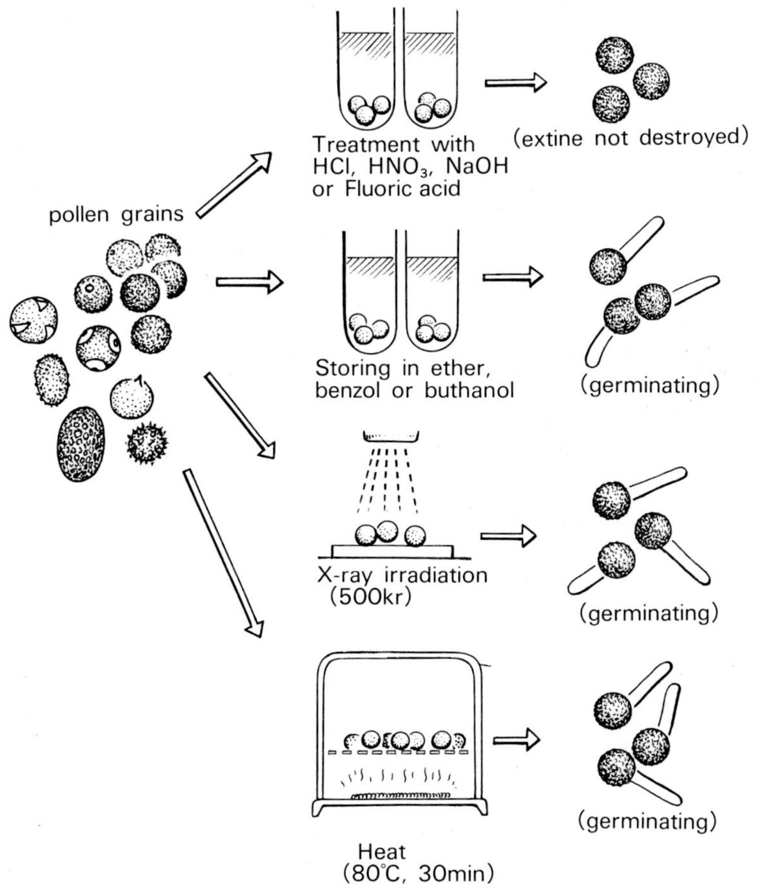

Fig. 4.37 Durability of pollen exine.

4.14 Components of pollen

Like other living cells, pollen grains generated in young anthers contain much water. When they are released from the anthers, the water content rapidly decreases. For example, in corn pollen at the anthesis stage, about 50% of the components is water, but soon after release, the water content in the grains becomes less than 10%.

Chemical components of pollen vary among species, but generally, 13–37% is carbohydrates, 11–20% proteins, 1–4% lipids, and 1–2% inorganic compounds.

As for atomic constitution, nitrogen makes up the largest portion after carbon, oxygen and hydrogen. The other major atoms in descending order of amount are: potassium, sulfur, phosphorous, calcium, magnesium and sodium. Slight amounts of boron, aluminum, copper, iron, cobalt, manganese and zinc are included in pollen grains. Fig. 4.39 shows the distribution patterns of the major elements found in lily pollen examined by X-ray microanalysis, indicating predominance of magnesium in the cytoplasm, calcium in the inner wall, and sulfate in the nucleus.

The kinds of protein found in pollen can be analyzed by electrophoresis and other means. Fig. 4.38 shows an immunological method of protein analysis between two different species, where an extract of pollen of one species is injected into a rabbit to obtain the specific antibody supplied for cross-reacting with protein from the other species. If the same number of precipitation bands appears between the antibody and the two extracts, the two species have common proteins.

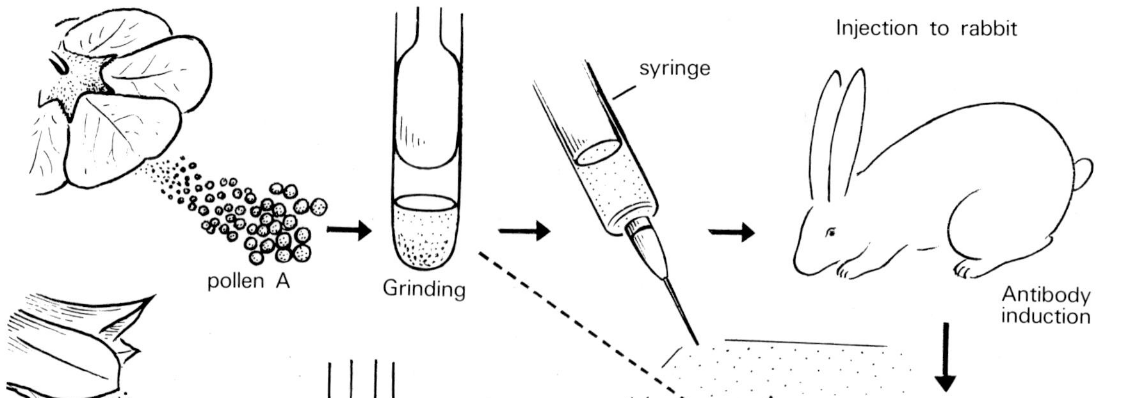

Fig. 4.38 Immunological examination of protein quality in pollen of two species. The antibodies induced in the serum of rabbit after injecting the extract from pollen of a species interact to aggregate on agar plate with extract from the same species.

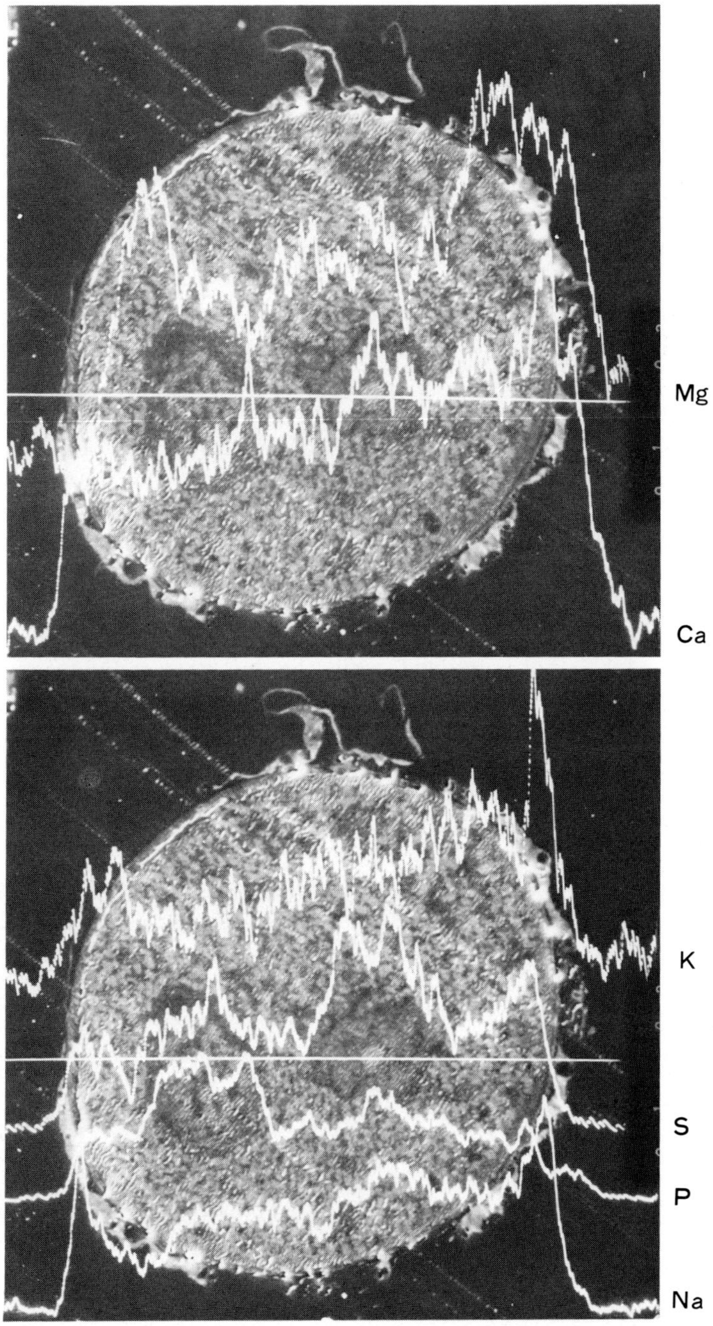

Fig. 4.39 X-ray microanalysis for distributions of ions in a pollen grain of Easter lily.
Top: Scanning for magnesium and calcium along line. Bottom: Scanning for potassium, sulfur, phosphorus and sodium along the line A–B. ×900.

5. Genetics of Pollen

Since pollen is the product of meiosis (reductional nuclear division), its genetic phase is haploid, and it functions as the male gamete through which genetic information is transmitted to the offspring. In other words, genetic information doubled in a parent plant is segregated into simple phases in each pollen grain. This is one of the basic concepts of modern genetics, known as allelic segregation, which was first proposed by Gregory Mendel, the father of genetics, in 1865. Therefore the character of pollen is the direct expression of an allelic gene in higher plants, and segregation seen among pollen produced from one plant is the allelic segregation of genetic information. However, since pollen is extraordinarily specified with a generally short life span, analysis of pollen genetics is relatively scarce.

5.1 Genetic segregation in pollen

One good example of allelic segregation in pollen is the waxy gene. It is known that certain varieties of several plants, especially cereals, produce waxy (starchy) seed grains; these include rice, corn, sorghum, barley and buckwheat. From waxy grains, cakes, cookies, dumplings, etc., are made utilizing the sticky and sweet characteristics of the starches. Whether a grain is waxy or non-waxy is controlled by a single gene, and the latter gene is dominant over the former.

In the waxy plant, amyropectin is the main chemical component of starch, and amyrose, the other starch component, is not contained, while the non-waxy individual has both amyrose and amyropectin (Fig. 5.1). Amyropectin is responsible for the sticky characteristic of the grain. Production of amyrose is controlled by one gene, designated as *wx*; the genotype of the waxy individual is *wx*/*wx*, and the non-waxy (normal) is wx^+/wx or wx^+/wx^+. This character is also expressed in pollen by the types of starch granules. The two can be differenciated by KI staining; amyrose stains pink, while the mixture of amyrose and amyropectin stains blue.

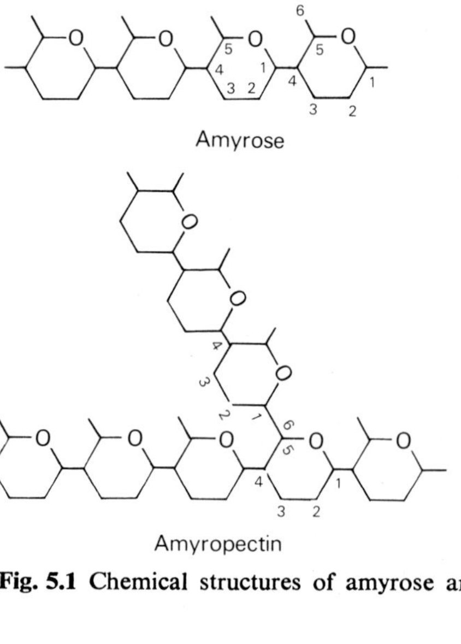

Fig. 5.1 Chemical structures of amyrose and amyropectin contained in starch of waxy pollen grains.

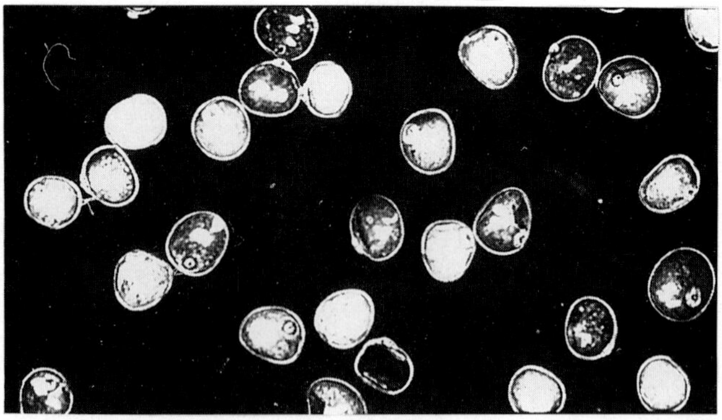

Fig. 5.2 Pollen grains of F_1 hybrid between waxy and non-waxy strains in corn, showing 1:1 segregation stainable with KI solution.

An F_1 hybrid of corn between waxy and non-waxy (wx/wx^+) produced two types of pollen grains one showing pink, and the other blue (Fig. 5.2). A 1:1 segregation is detected between them. This is visual evidence of allelic segregation of a gene.

5.2 Pollen sterility

Other genetic characteristic appearing in pollen is sterility. Sterile pollen (non-functional pollen) are produced under restricted environmental conditions such as cold or high temperatures, and nutrient shortage. But sterility is also caused by genetic events. This includes genic control (pollen sterility gene, pollen killer gene, gametocidal gene, etc.), chromosomal aberration, non-homology of genomes and cytoplasmic abnormality. For example, if a pair of homologous chromosomes (one from the maternal, and one from the paternal) fail to pair at meiosis, like chromosomal translocation, the products include sterile (abnormal, or vacant in most of cases) pollen (Fig. 5.3) because of unequal distribution of the chromosome set. If a tetrad (four grains from a single meiocyte) does not separate each other, the segregation of sterile and fertile components are easily detected in a unit (Fig. 5.4). These attached grains can be obtained if they are sampled at a relatively early stage of anthesis, or if the spikes or buds are treated with low temperature.

Generally, F_1 hybrids between different species produce sterile pollen. This sterility results from chromosome unpairing in meiosis due to the lack of chromosomal homology, causing unbalanced or abnormal distribution of chromosomes in the reductional division. In other words, genetic homology between two species is measured by analyzing the degree of chromosome pairing in an interspecific hybrid. The production of sterile pollen in interspecific hybrids is one of the mechanisms of species isolation.

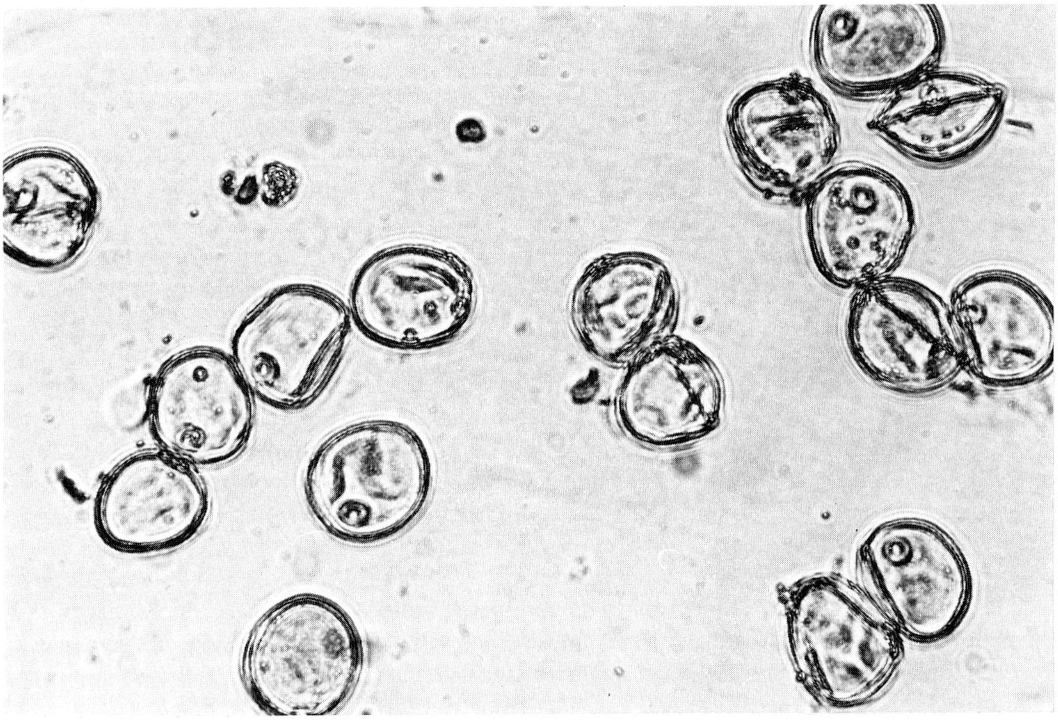

Fig. 5.3 Complete sterile pollen in alloplasmic wheat (*Triticum aestivum* having cytoplasm of *T. timopheevi*).

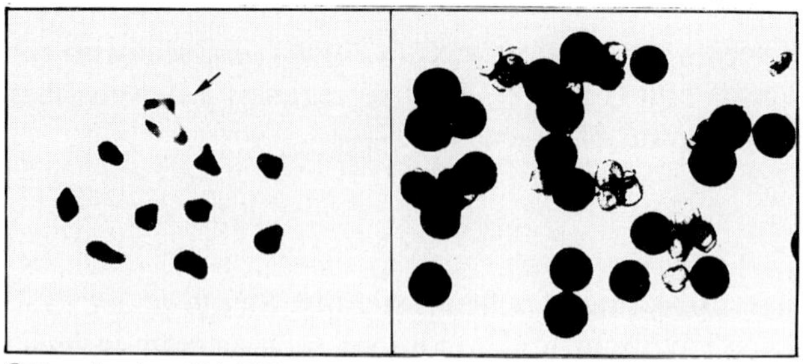

④ $^{+}9_{II}$configuration at MI in PMC's (±8000)   Pollen tetrads (×400)

Frequency of 5 types of pollen tetrads in this interchange heterozygote

Type	1 ○○○○	2 ○○○×	3 ○○××	4 ○×××	5 ××××	Total	Percentage of type3 tetrad (c.o. tetrad)	Type1 : type5 (Altern : Adjacent) segregation	Pollen fertility (%)
Frequency	66	2	145	8	25	246	58.9	2.6 : 1	57.7

○: normal pollen grain. ×: empty pollen grain

Fig. 5.4 Tetrad analysis of pollen sterility in an interchange heterozygote plant of watermelon. Ring-of-four formation (arrow) in meiotic chromosome pairing causes partial pollen sterility. Since one tetrad is the product of one PMC, the type of tetrad represents the pattern of unbalanced distribution of the interchanged chromosomes.

5.3 Cytoplasmic male sterility

It has been demonstrated that mitochondoria and chloroplasts carry genetic information, that is, they have their own DNAs. In the process of fertilization of the most of higher plant, however, mitochondria and chloroplasts are not transmitted through pollen to the offspring. From the male gamete, the generative nucleus contributes only to fertilization; it does not transmit cytoplasmic factors. Therefore, the genetic information in mitochondria or chloroplast DNA are maternal inferited.

The cytoplasmic genes carried in mitochondria and chloroplast DNAs function in cytoplasmic organelle biosynthesis. At the same time, some of these genes interact with nuclear genes to regulate cell metabolism. In plants, it has been demonstrated that genetic factors of cytoplasm affect pollen genesis because reciprocal crosses show differential pollen fertility in cetain hybrids. F_1 from the cross of A(♀) × B(♂) produces fertile pollen, while B(♀) × A(♂) produces sterile pollen. Since cytoplasmic factors are transmitted only through female parents, it is possible to substitute cytoplasm of one species with that of another through successive backcrossing using a given species as the recurrent male parent between A and B species, as illustrated in Fig. 5.5.

In wheat (*Triticum aestivum*), cytoplasmic substitution lines (alloplasmic wheat strains) have been established by the above-mentioned method, using various *Aegilops* species (wild relatives of wheat) as the cytoplasmic donors. Interestingly, most of these lines show male sterility, in addition to other abnormalities such as lack of vigor, twin formation, and/or delayed maturation. Male sterility includes incomplete developement of anthers (Fig. 5.6), pollen sterility (Fig. 5.7), and pistiloidy (abnormal development of stamens appearing on pistils) (Fig. 5.8). Female organs in these lines do not show any abnormality.

These phenotypes depend on the combinations between kind of cytoplasm and nuclear genotypes. Genetic analyses have revealed that there is one or more nuclear genes which restore these abnormalities (male-sterility restoring gene) for each combination. Although the detailed mechanisms of cytoplasmic male sterility have not been clarified, this phenomenon is widely applied in agriculture. Hybrids between different varieties sometimes exhibit hybrid heterosis (in which offspring are stronger than the parent varieties), which is beneficial for increasing yield of crop grains and vegetables. However, artificial hybridization (removal of stamens and pollination with pollen from another variety using tweezers) is not practical. If a cytoplasmic male sterile variety is grown side by side with the male fertile (normal) line of the other variety, the seeds produced on the male sterile plants are of the F_1 hybrid type, which are used in agricultural production to take advantage of hybrid heterosis. This system (known as CMS hybrid seed production) is used in the production of corn, rice, many vegetables and horticultural plants.

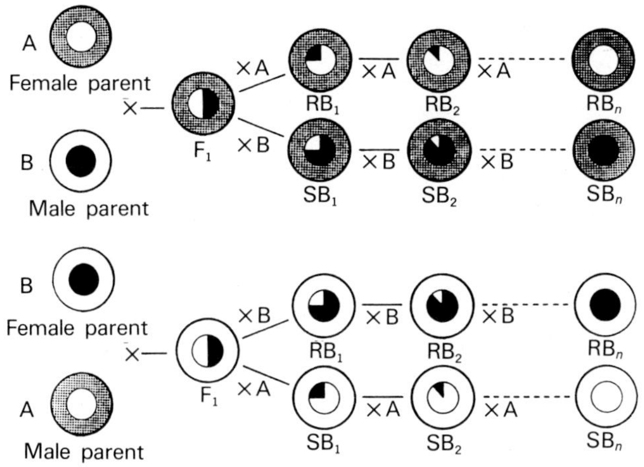

Fig. 5.5 Procedure for establishing nucleus-cytoplasm substitution lines by successive backcrossing between two species.

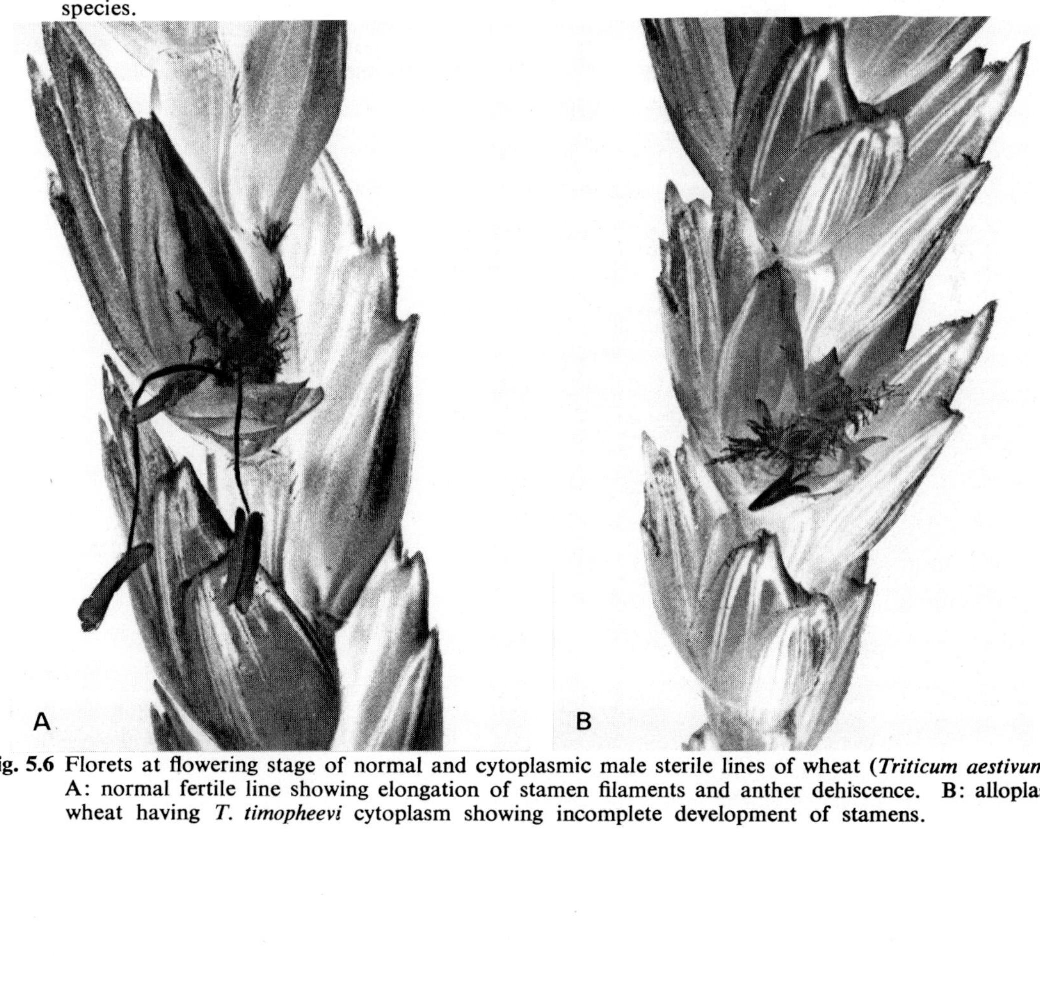

Fig. 5.6 Florets at flowering stage of normal and cytoplasmic male sterile lines of wheat (*Triticum aestivum*). **A**: normal fertile line showing elongation of stamen filaments and anther dehiscence. **B**: alloplasmic wheat having *T. timopheevi* cytoplasm showing incomplete development of stamens.

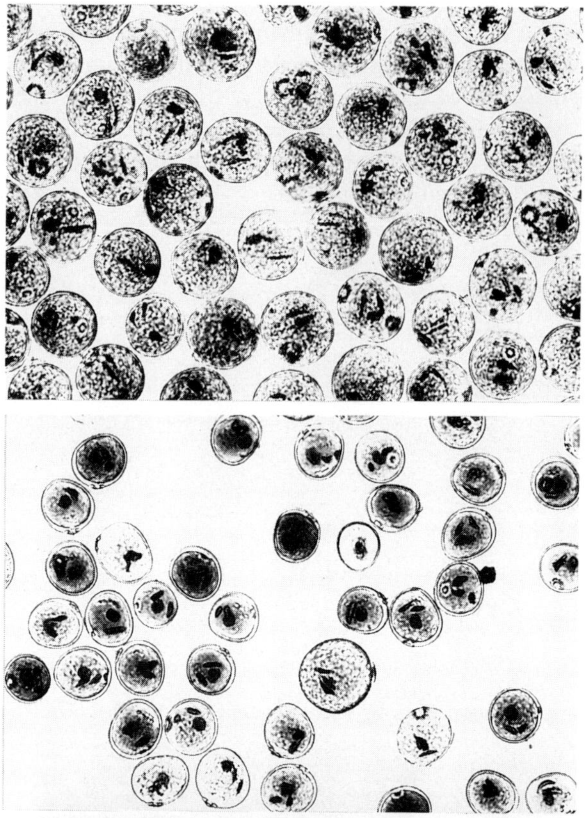

Fig. 5.7 Pollen grains of complete fertility from normal line (top), and partial sterility from alloplasmic line (bottom) in wheat. Most cytoplasmic substitution does not affect function of the female organs.

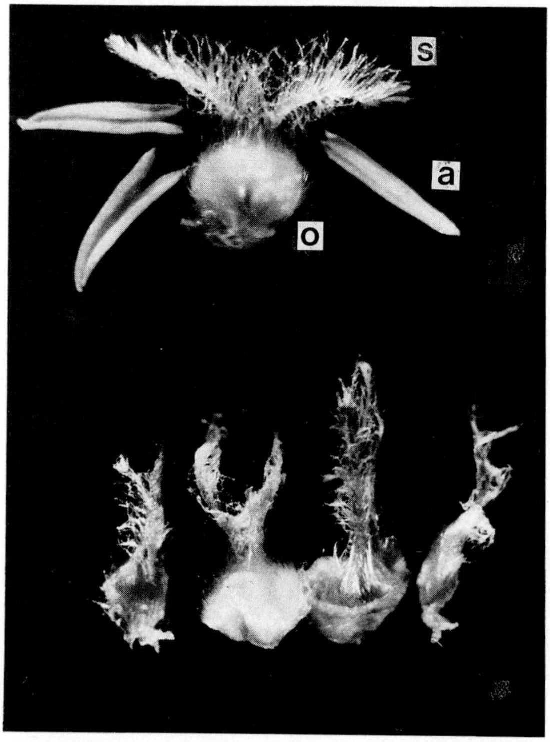

Fig. 5.8 Pistiloidy of wheat stamens induced by cytoplasm substitution
Top: normal line showing full development of ovule (o), stigma (s) and three anthers (a). Bottom: cytoplasm substitution line with *Aegilops crassa* having malformed stamens of pistil-like structure (pistiloidy).

5.4 Natural polyploidization

About 70% of plant species are polyploid, *i.e.* having more than three sets of genomes. Polyploidy includes two categories, auto-polyploidy and allopolyploidy. The former is produced by the simple doubling of a chromosome set of a diploid species. The latter, having different genome sets, is achieved in two steps, that is, fertilization between two species, and chromosome doubling of the F_1 hybrid. Artifically, a plant with doubled chromosome (artificial polyploidy production) is obtained by colchicine treatment on the apical buds or meristematic tiller of a diploid plant. But how does natural polyploidization occur in plant evolution?

In nature, interspecific hybridization often occurs between related species of plants. However, the F_1 hybrids are usually sterile because of failure of chromosome synapsis in meiosis between two species. This is an isolation mechanism of plant species.

Bread wheat (*Triticum aestivum*) has a chromosome number of $2n = 42$, with a genome (unit of chromosome set) constitution of AABBDD, indicating allopolyploidy of three genomes (hexaploid wheat). On the other hand, wheat for spaghetti (*Triticum durum*, $2n = 28$) has a genome constitution of AABB (tetraploid wheat). It has been discovered that one type of tetraploid wheat hybridized with a wild diploid weed, *Aegilops squarrosa* ($2n = 14$; DD), to produce the AABBDD hexaploid wheat in evolution (Fig 5.9). This was demonstrated by artifiial hybridization between the two

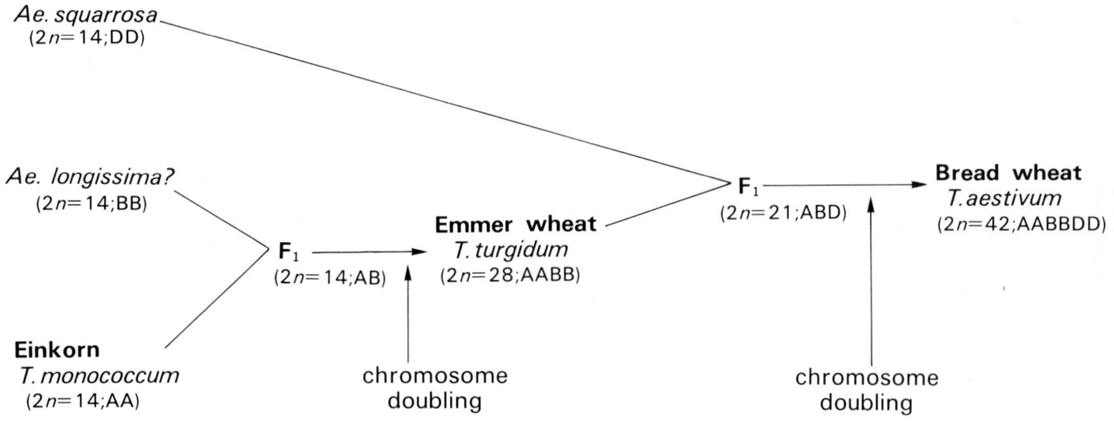

Fig. 5.9. Evolutional pathway of bread wheat.
Inter-specific hybridization and chromosome doubling are involved to establish allopolyploid species.

species and colchicine treatment of the F_1 hybrid to obtain a synthesized hexaploid wheat. However, when certain strains were used as the parent lines for hybridization, fertile F_1 producing self-pollinated F_2 seeds were obtained without colchicine treatment (Fig. 5.10). The F_1 plant had a chromosome constitution of $2n = 21$, while the F_2 plants had $2n = 42$. This indicates that natural polyploidization occurred in this hybrid, producing a natural hexaploid wheat. In Fig. 5.12. cytological events during meiosis of pollen mother cells (PMCs) are presented in the three cases of interspecific hybrids. The fertile F_1 showed complete asynapsis (inhibition of chromosome pairing) in meiosis, while the other combinations had some paired chromosomes. Furthermore, the fertile F_1 had only one nuclear division in meiosis resulting in two microspores as the meiotic

Fig. 5.10 Spikes of synthesized wheat and its parents.
A: *Aegilops squarrosa* ($2n=14$; DD) used as the female parent.
B: *Triticum durum* ($2n=28$; AABB) used as the male parent.
C: Fertile F_1 having chromosome number of $2n=21$.
D: F_2 plant obtained from selfing of F_1 ($2n=42$; AABBDD).

product, while the normal line or the sterile hybrids had two divisions producing four products. This is the process of unreduced gamete formation, which is illustrated in Fig. 5.11. The unreduced gamete formation must occur in the female side, and self-pollination produce doubled chromosome constitution (AABBDD production). Since this occurred only in certain combinations, the gene(s) of asynapsis should be involved in the mechanism.

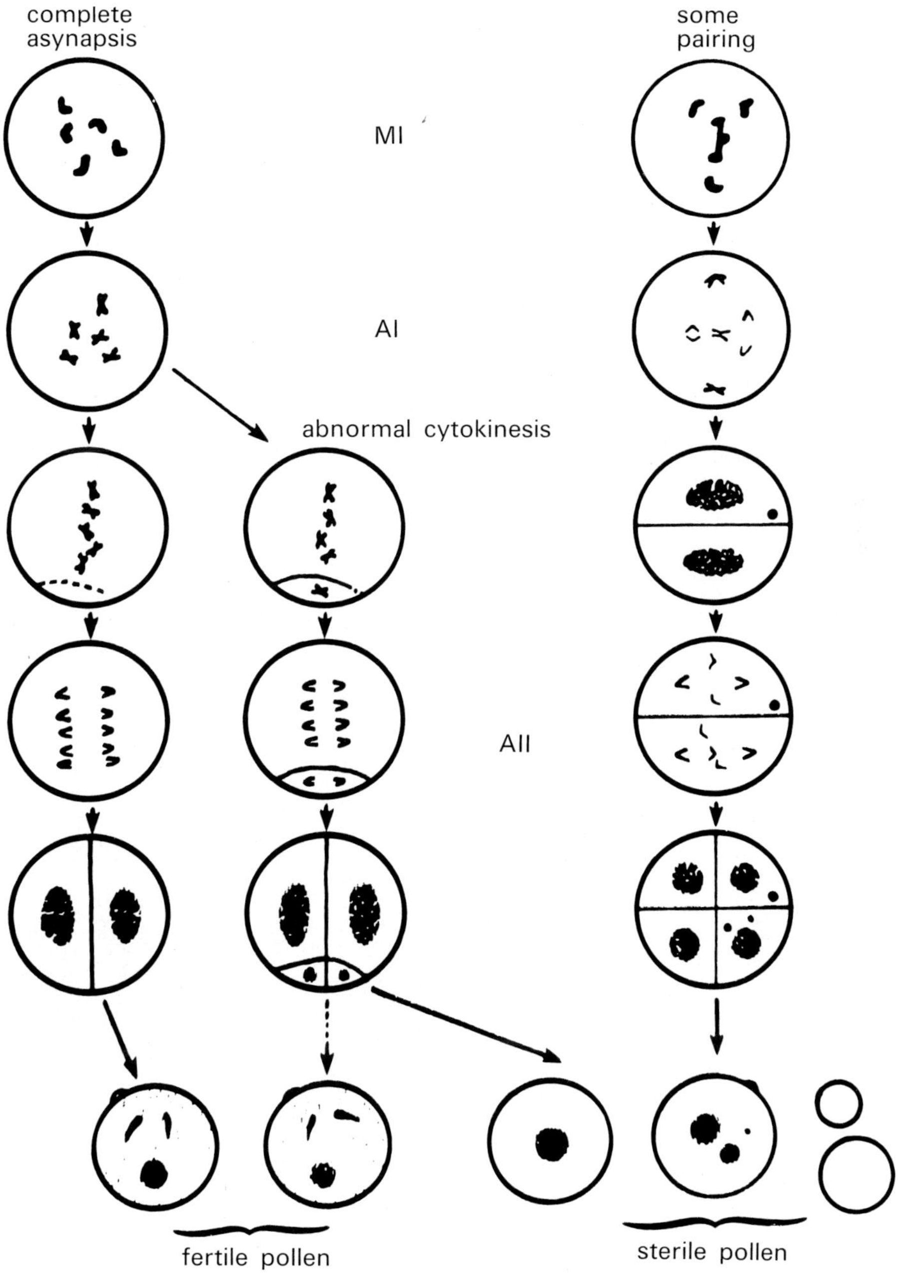

Fig. 5.11 Possible scheme of miotic processes for producing fertile and sterile pollen in interspecific hybrids. Complete asynapsis in PMC's initiates meiotic restitution at division I and equational division at division II, resulting in the formation of symmetrical dyads and fertile pollen grains. Abnormal cytokinesis may or may not cause meiotic nonreduction. Presence of some pairing brings two cell divisions at meiosis, resulting in sterile pollen porudction.

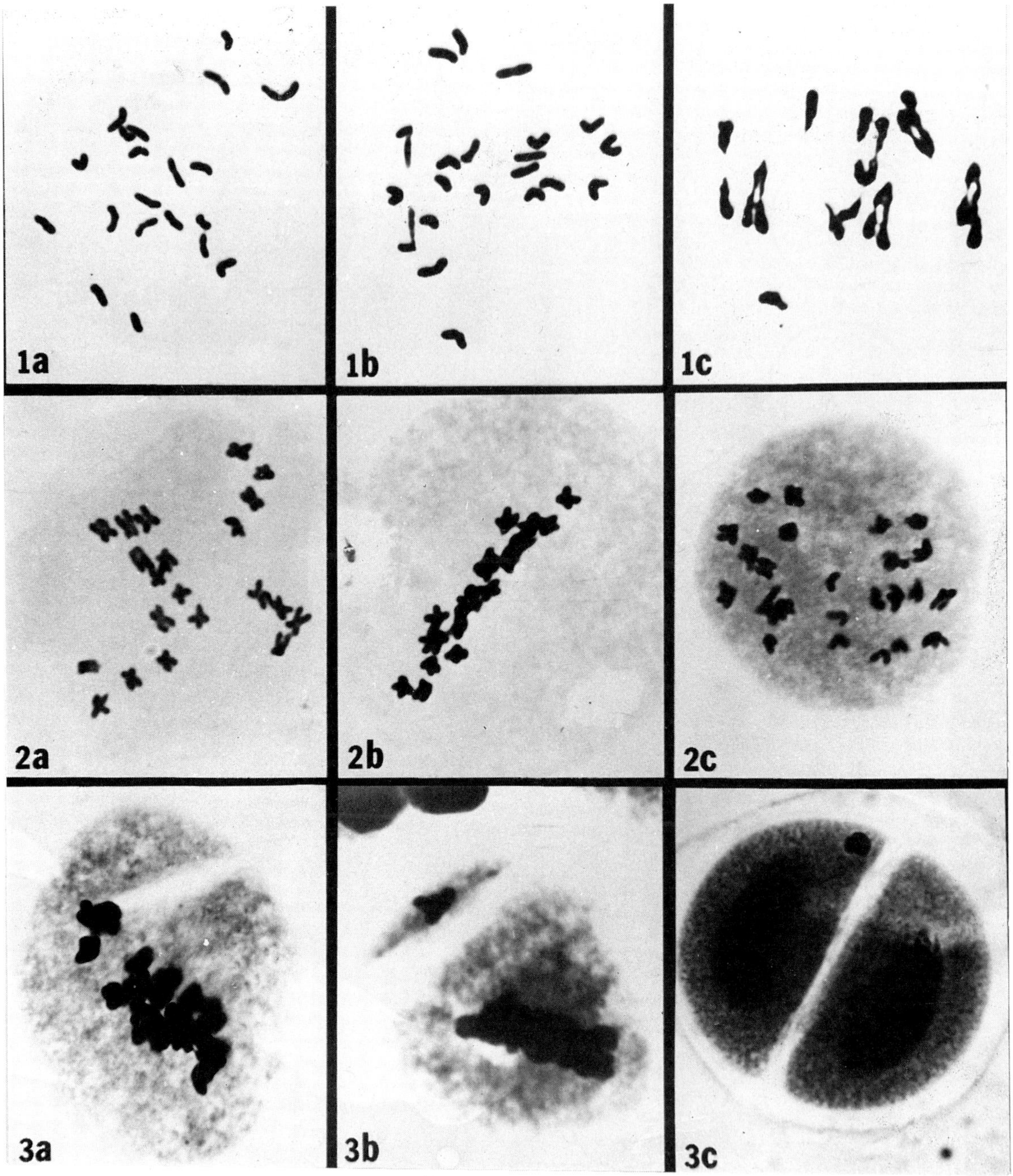

Fig. 5.12 Meiotic processes in PMC's of fertile, partial fertile and sterile F_1 hybrids.
a: *Ae. heldreichii-T. durum* F_1. b: *Ae. ligustica-T. durum* F_1. c: *Ae. spletoides-T. durum* F_1.
1: Chromosome pairing at MI; complete asynapsis (la), $1_{II}+19_{I}$ (1b), and $4_{III}+4_{II}+1_{I}$ (1b). 2: Late MI-AI; PMC's of 2a and 2b may be in restituted state. In 2c, paired chromosomes separate reductionally,

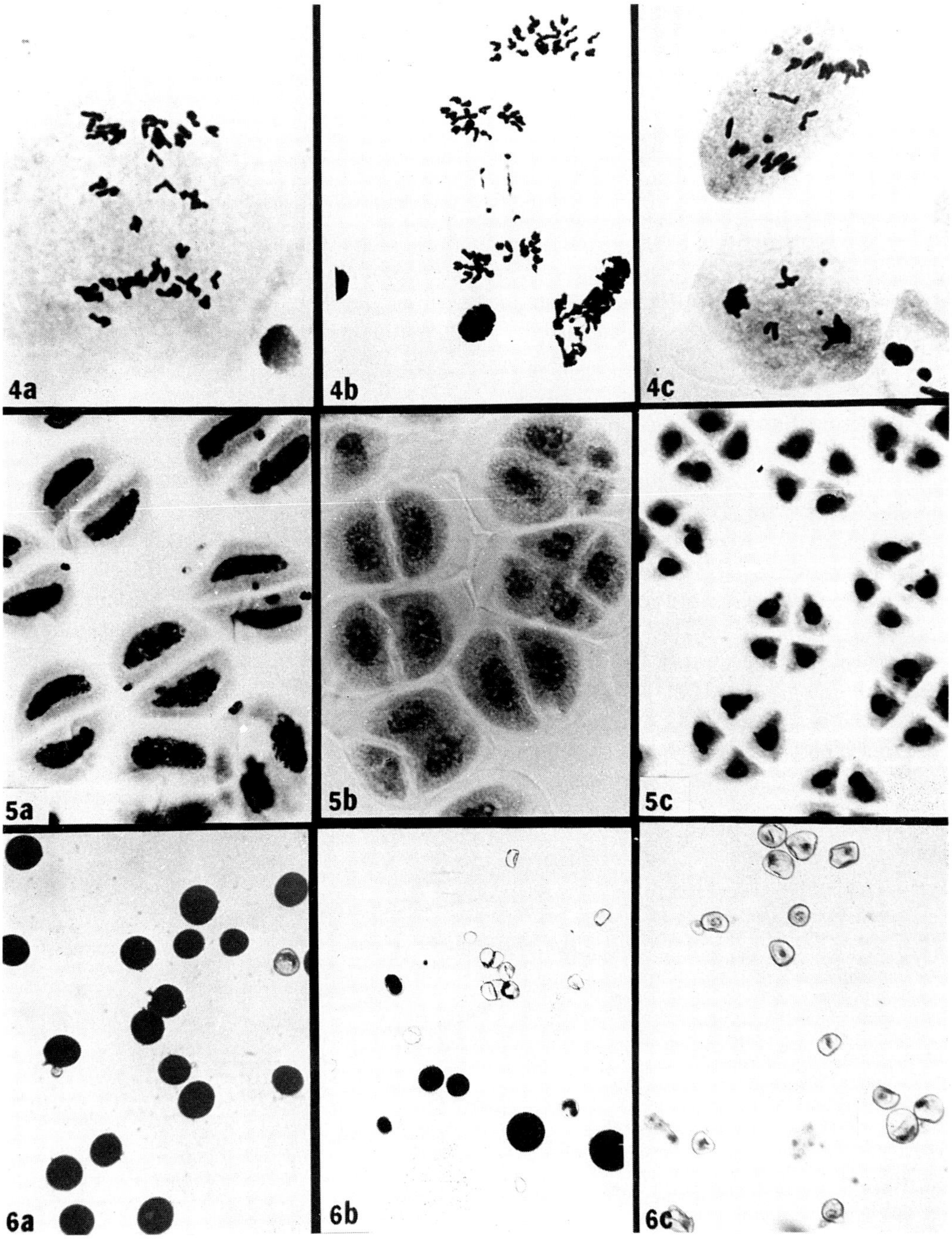

while univalent chromosomes divide equationaly. 3: Abnormal cytokinesis (3a and 3b), and dyad (3c). 4: AII: In 4a and 4b, chromosomes divide equationally. In 4c, chromatids originated from univalent chromosomes at MI segregate randomly. 5: "Tetrad" stage: Formation of symmetrical dyads (5a and 5b), and tetrads with micronuclei (5c). 6: Pollen at anthesis stage.

6. Pollen in Air and Pollinosis

There are numerous species of anemophilous plants fertilized with the help of wind, so it is natural for many pollen grains to be in the air (Fig. 6.2). These are atmospheric pollen, and their quantities increase during the flowering seasons of anemophilous plants, causing pollinosis, or pollen allergy, in numerous individuals with accompanying symptoms of rhinitis or asthma. Recently, atmospheric pollen is being extensively collected and analyzed in order to identify antigens for pollen allergy.

6.1 Collection of atmospheric pollen

Figure 6.1 shows a method for collecting atmospheric pollen grains, illustrating a standard pollen collection system (Durham's pollen collecter; stainless discs 23 cm in diameter upon which glass slides smeared with vaseline are placed, are set up 9 cm from ground level). The slides are microscopically examined at 24 hours intervals to analyze the amount and kind of pollen adhering to the slide. For analysis, Calbela's solution (5 ml glycerin, 10 ml 95% ethanol, 15 ml distilled water and two drops of saturated fuchsine) should be mounted over the sample. Examining the daily pattern of pollen appearance over one year, the varieties of pollen and their amounts can be calculated by counting the number of pollen grains per 1 cm^2 on the slides.

Other methods for collecting atmospheric pollen are glass slides smeared with vaseline being revolved in a high place or attached to the front of a helicoptor to collect pollen grains in the sky (Fig. 6.3).

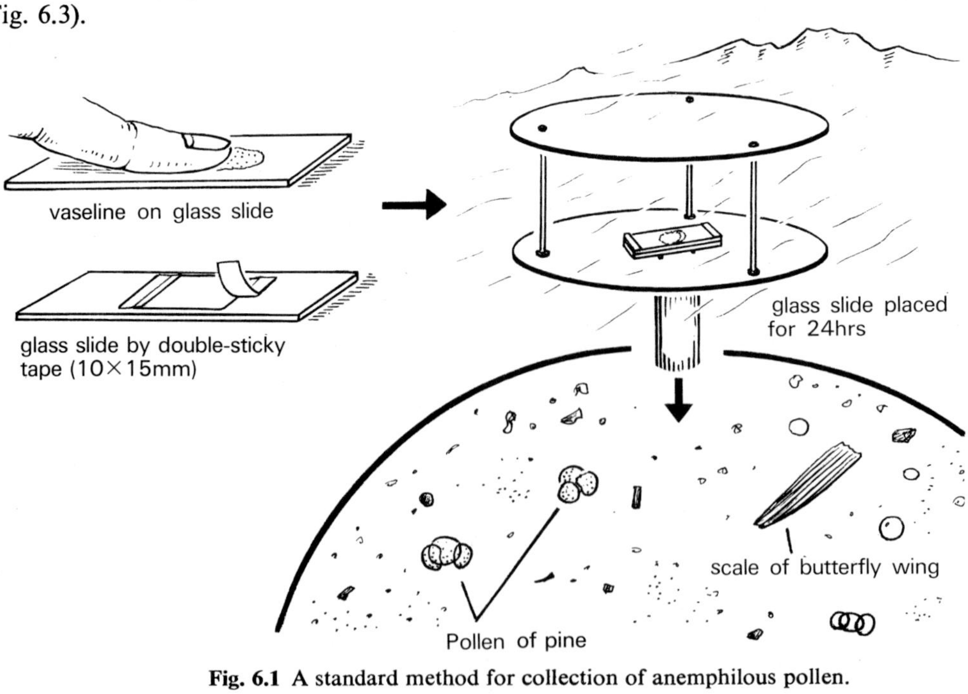

Fig. 6.1 A standard method for collection of anemphilous pollen.

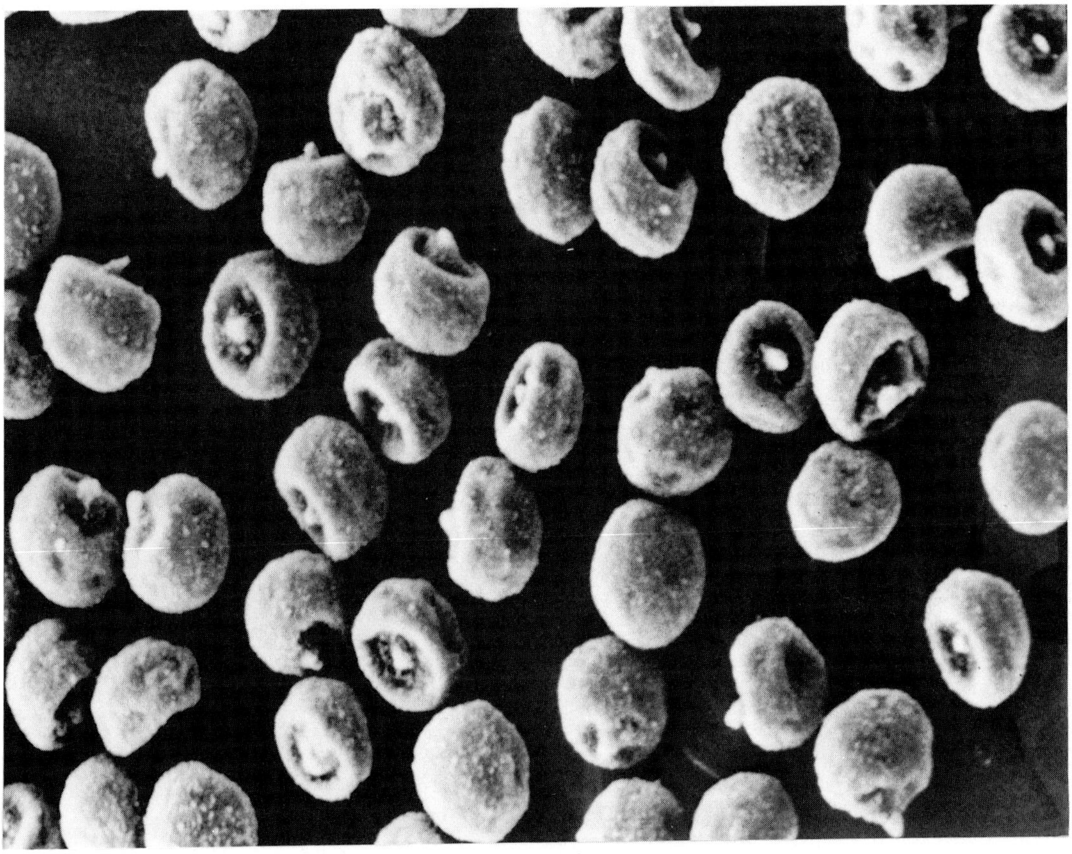

Fig. 6.2 Pollen grains of Japanese red cedar (*Cryptomeria japonica*) in air (×770).

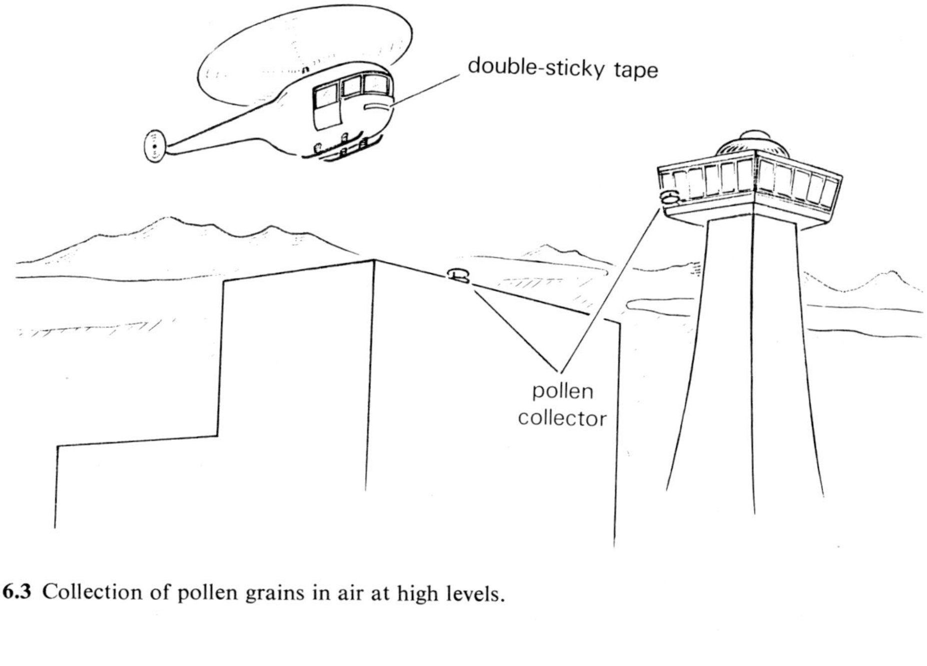

Fig. 6.3 Collection of pollen grains in air at high levels.

6.2 Seasonal changes of atmospheric pollen

In the United States, three seasons are listed according to the dominant types of atmospheric pollen, *i.e.* the tree season (pollen from trees; March–May), the grass season (pollen of *Grammineae*, May–July), and the weed season [pollen from orchard grass (*Dachylis glomerata*), and ragweed (*Ambrosia artemisiaefolia*)]. Japan also has these seasons, although the third is not so distinctive except in the Hokkaido area where orchard grass and timothy are cultivated as pastures. Pollen of *Grammineae* species drift in the air for relatively long periods in Japan, especially from May to June, and August to September (Fig. 6.4).

Pollen types, seasons and amounts vary between localities and between years. The graphs in Fig. 6.6 were drawn from observations made in Kanagawa Prefecture in Japan. Although most atmospheric pollen comes from anemophilous flowers, pollen of entomophilous, or insect-pollinated, flowers are sometimes caught on slides; these include goldenrod (*Solidago altissima*), pear (*Pyrum serotina*), and tobacco (*Nicotiana tabacum*) (Fig. 6.5).

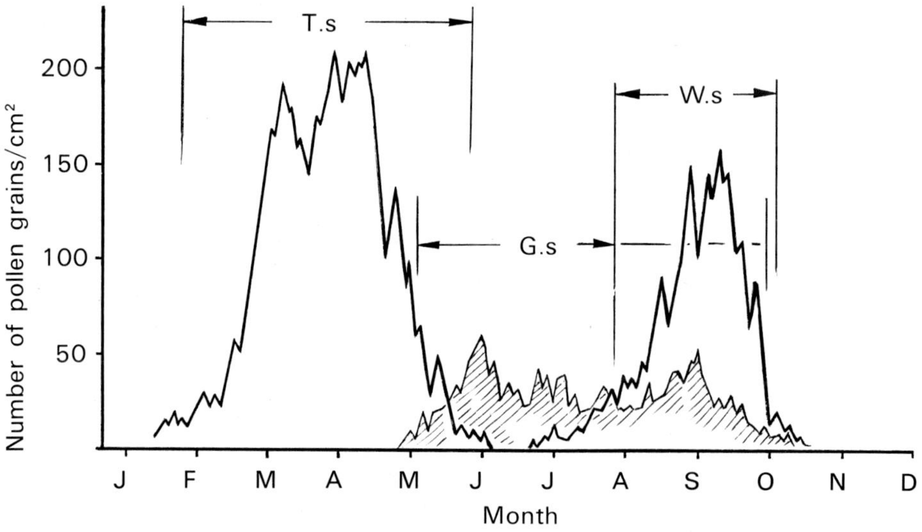

Fig. 6.4 Seasonal changes of anemophilous pollen.
T.s: Tree season. G.s: Grass season. W.s: Weed season.

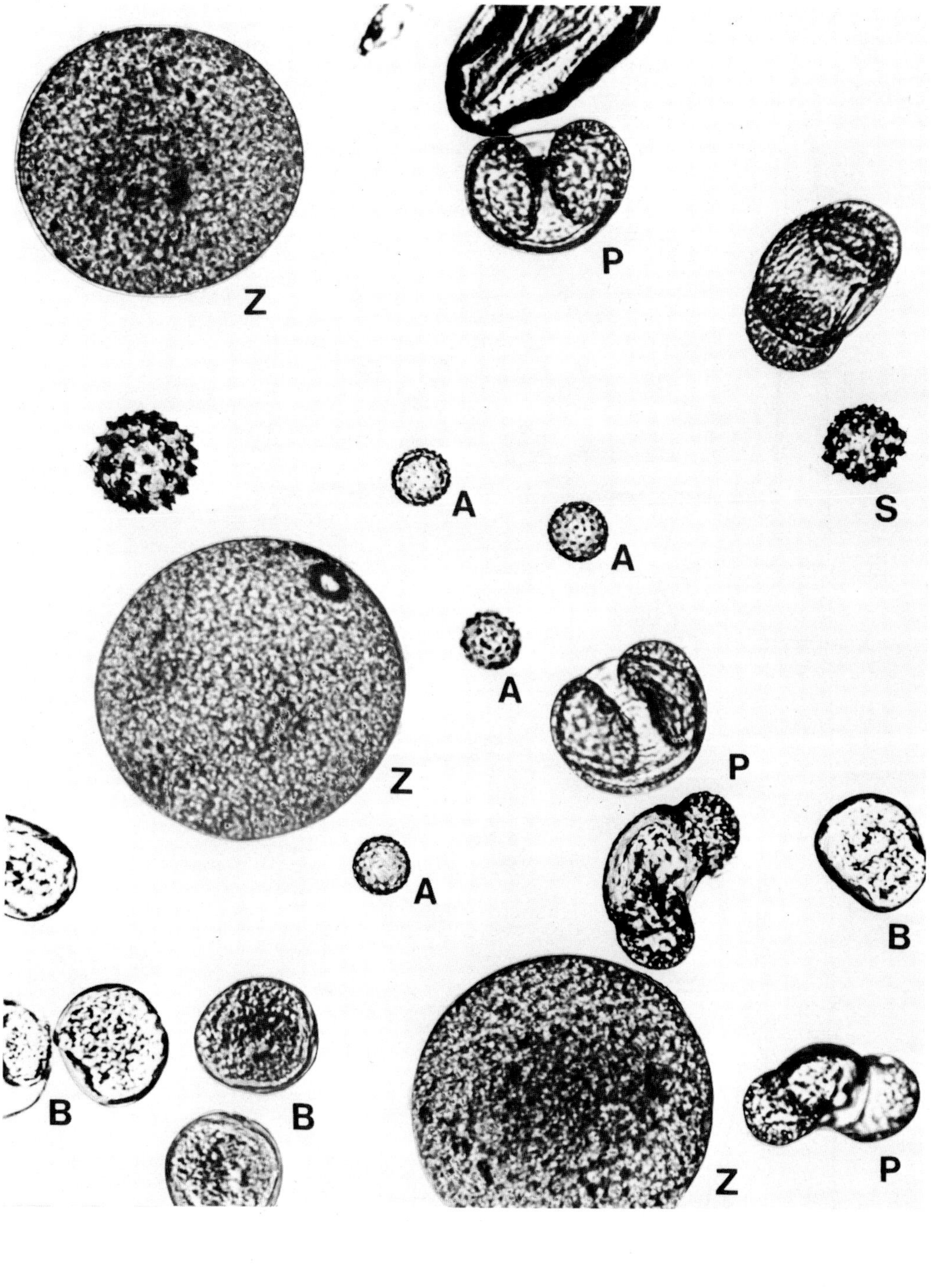

Fig. 6.5 Various kinds of pollen grains collected from air.
B: grass (*Gramineae* sp.). A: ragweed (*Ambrosia artemisiaefolia*). P: pine (*Pinus* sp.). S: goldenrod (*Solidogo altissima*). Z: corn (*Zea mays*).

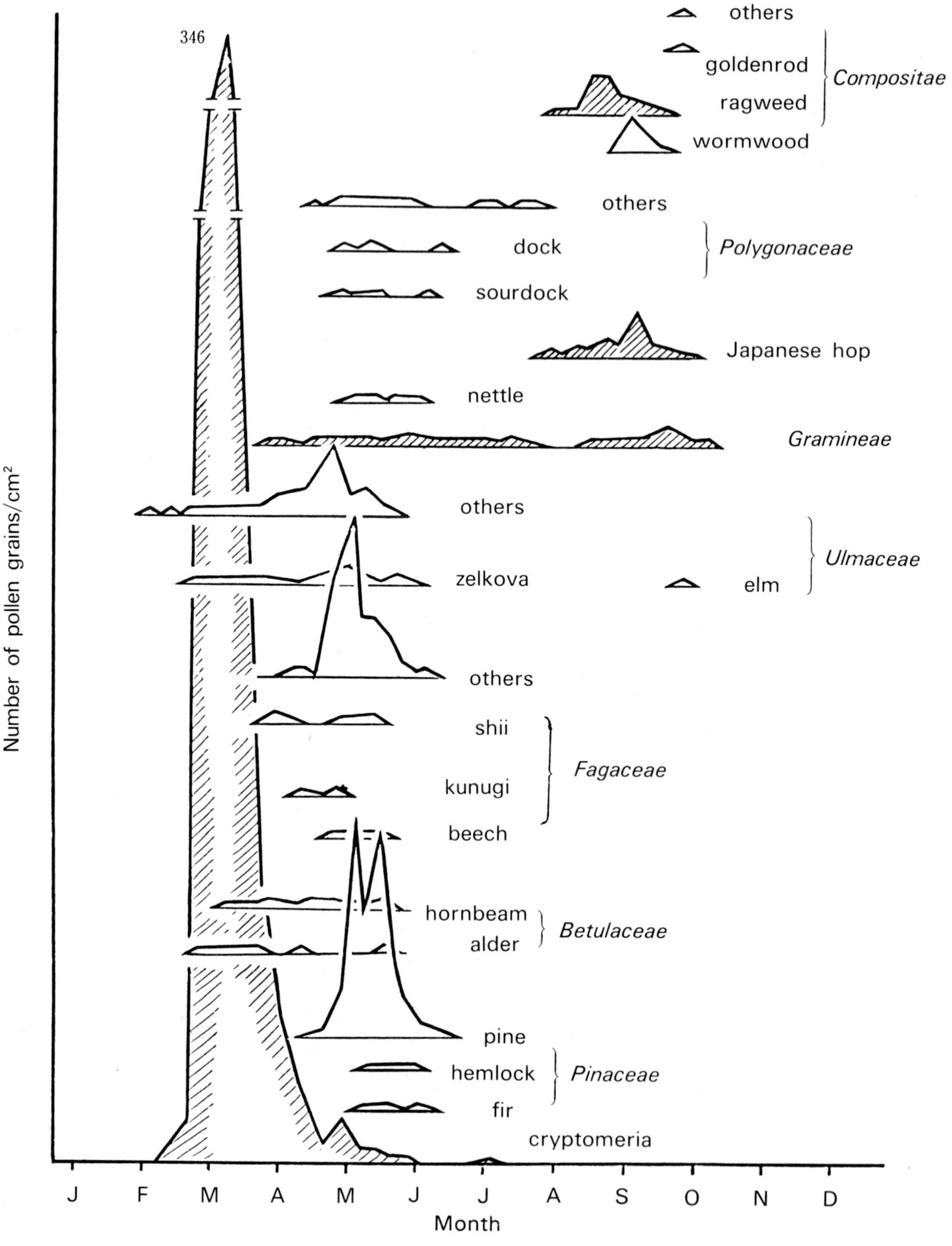

Fig. 6.6 Types and amounts of pollen observed over one year in Kanagawa prefecture, Japan. (Recorded by Dr. T. Shida).

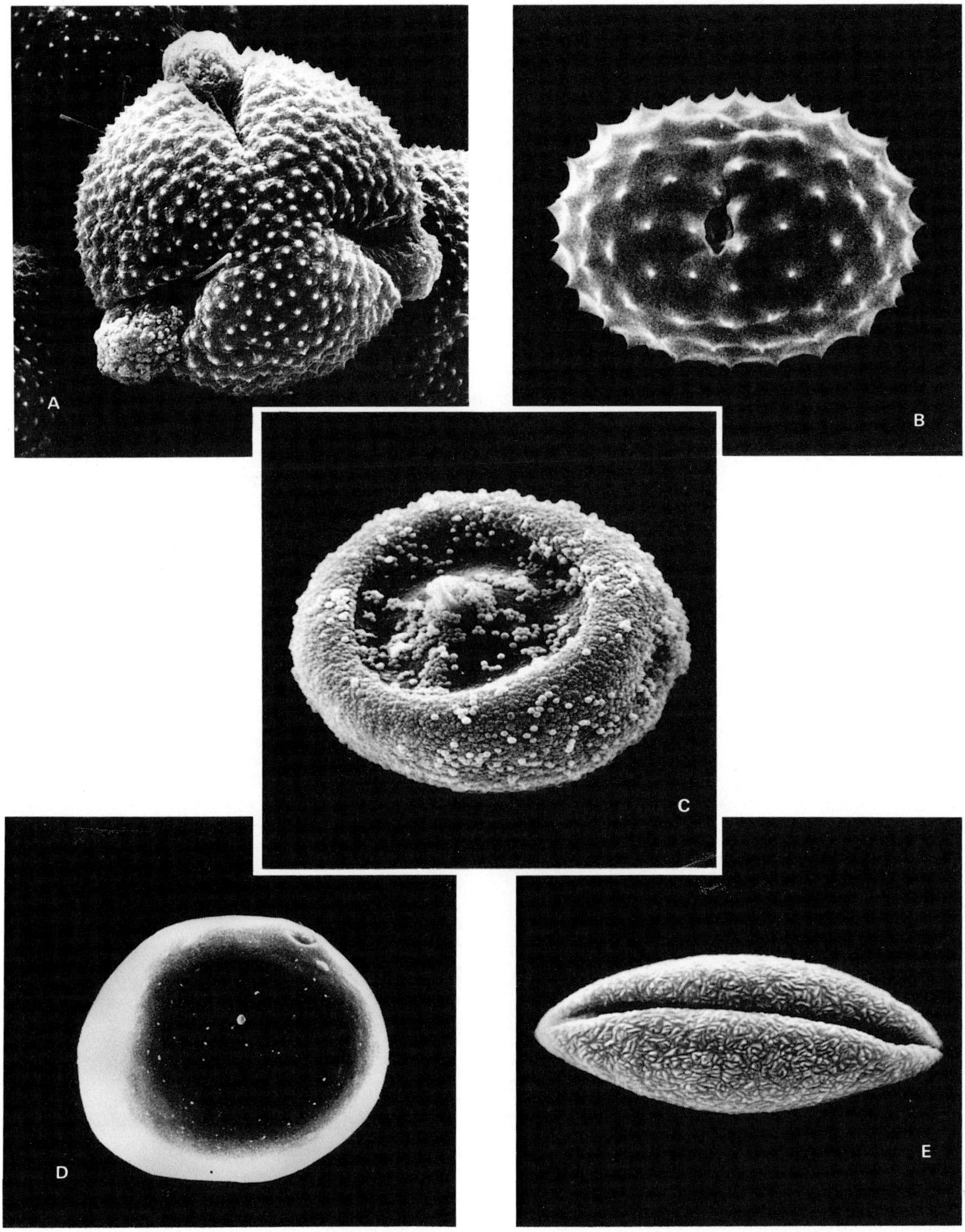

Fig. 6.7 Pollen grains which cause widespread pollinosis.
A: common mugwort (*Artemisia vulgaris*) (×3200); B: bitterweed (*Ambrosia artemisiaefolia*) (×2200). C: Japanese red cedar (*Cryptomeria japonica*) (×1700); D: sweet scented oleander (*Nerium indicum*) (×1200); E: maidenhair (*Ginkgo biloba*) (×1600).

6.3 Pollinosis

When a person is allergic to pollen, he or she suffers from rhinitis or asthma during a certain pollen season. If pollen grains adhere to the nasal mucous membrane sneezing and runny nose result, and in severe cases, a feeling of powerless or debility with headache or buzzing in the ears can occur.

Presently, one of every ten Japanese suffers from pollinosis. When extracts from pollen are injected into the epidermis of patients, positive skin reactions are recognized, indicating the production of immunoglobulin (antibodies) in serum. About ninety species have been identified thus far pollen stimulates antigen activity in humans; these include Japanese red cedar (*Cryptomeria japonica*), common mugwort (*Artemisa vulgaris*), rice (*Oryza sativa*), bitterweed (*Ambrosia artemisiaefolia*), birch (*Betula tauschii*), sweet-scented oleander (*Nerium indicum*), maidenhair (*Ginkgo biloba*), and chestnut (*Castanea crenata*).

Lymphocytes isolated from a pollinosis patient and cultured *in vitro*, the indicate the following differences in comparison with those from non-pollinosis individuals: There is no difference between the cultures at first, but when incubated for ten minutes the patient's lymphocytes are associated with pollen gains, one to two grains per lymphcyte on average, and more than ten in severe cases (Fig. 6.8 A,B). Extensive microvilli are identified in these lymphocytes attached to the grain wall (Fig. 6.9). The surface of the pollen grain changes dramatically—the entire surface of the pollen grain becomes porous and amorphous, sometimes with dissolution of its specific design (Fig. (Fig. 6.10). The above facts indicate specific interaction between lymphocytes and certain pollen, lymphocytes of a pollinosis patient attracting pollen with which to combine. The pollen grain is structurally changed upon attachment to lymphocytes. This is the first step of pollen allergy.

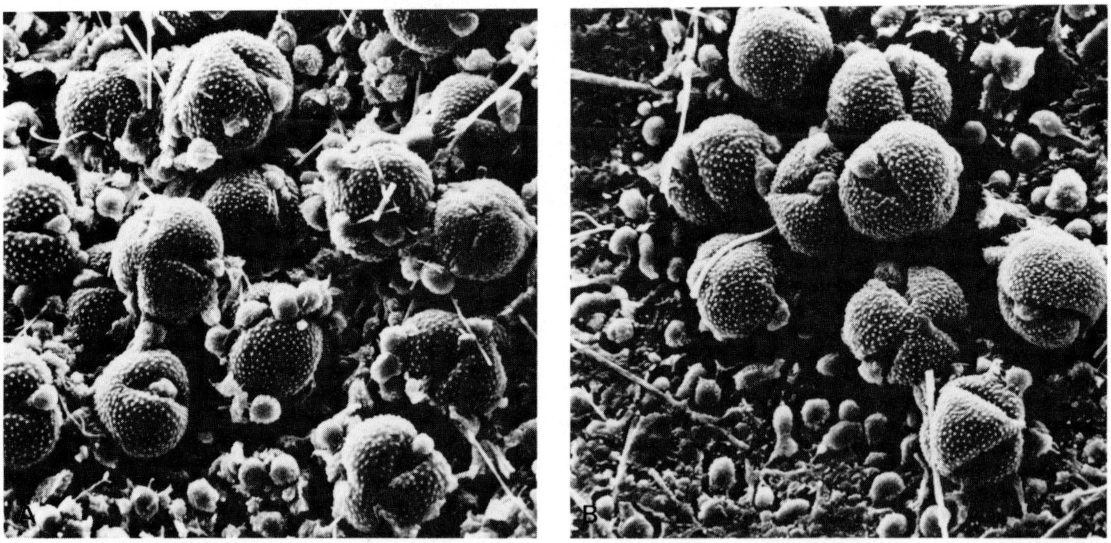

Fig. 6.8 Interaction of mugwort pollen grain with lymphocyte. A: Pollen grains attached to lymphocytes when pollen and lymphocyte from a pollinosis patient are incubated together. ×800. B: They do not interact in the case of non-pollinosis. ×800.

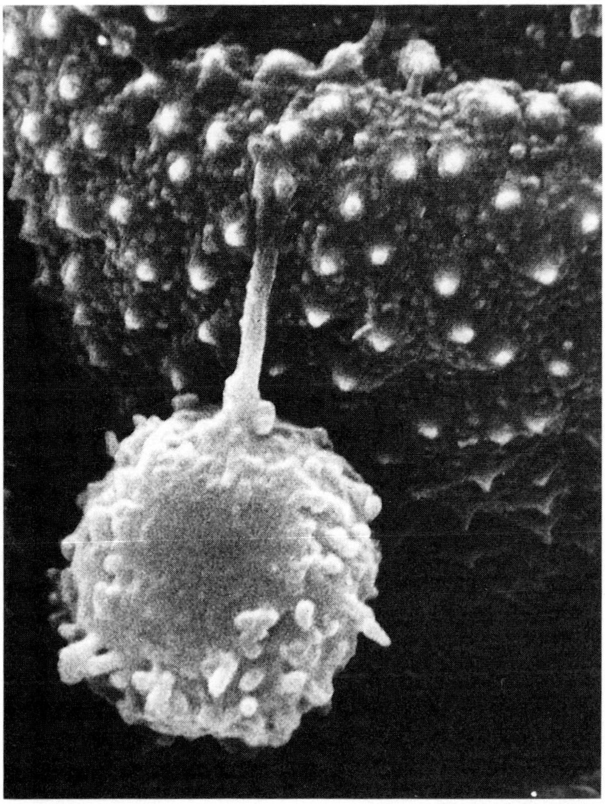

Fig. 6.9 SEM micrograph of the patient's lymphocyte, showing extensive microvilli. ×8400.

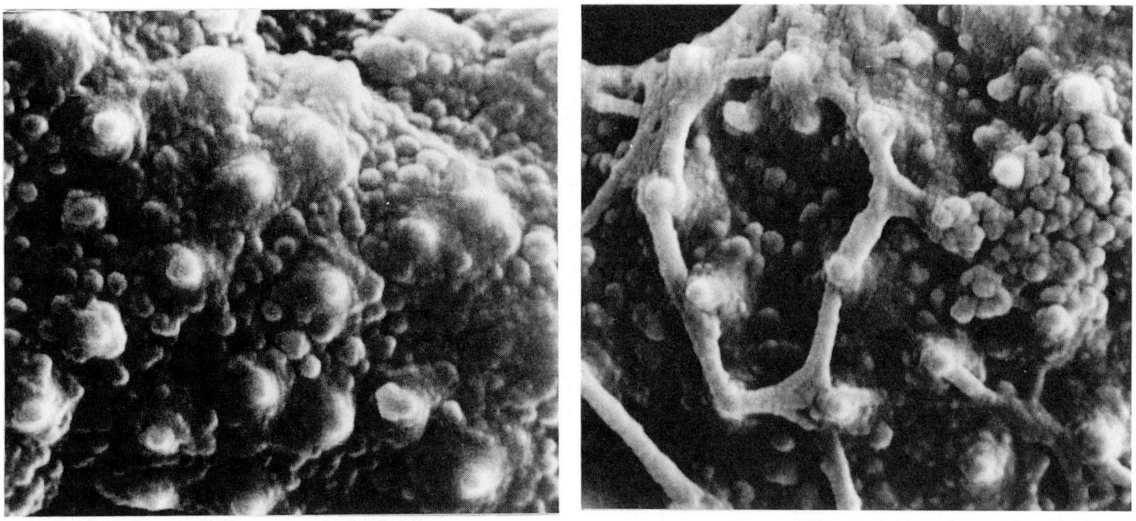

Fig. 6.10 Mature, intact exine surface of common mugwort pollen (*left*). ×15000. Exine surface, 10 min after incubation, showing dissolution of its specific sculpture (*right*). ×15000.

7. Pollen in Soil

After pollen is released from anthers, only a few grains eventually reach the stigmas of pistils to transmit genetic information to the next generation through fertilization. Most fail to meet partners for mating and fall to the ground to no purpose. Fallen grains are swept away by water to accumulate in soil. The grain material inside the intine (inner pollen wall) is destroyed within a short time in the soil, but the extine (outer pollen wall), which consists of a tough material called sporopollenin remains intact in soil for thousands of years; indeed, some have existed as long as a hundred thousand years (Fig. 7.2).

7.1 Pollen analysis

The term "pollen analysis," although it seems to indicate chemical analysis of pollen components, is used rather to signify identification and characterization of pollen grains in soil layers, which reflect environmental conditions of the earth at the time each layer was formed. Plant species propagated in the past can be determined by pollen analysis which examines the types and amounts of pollen grains in soils. The pollen analysis usually includes identification of spores and foraminiferas as well. Weather conditions of various periods can be determined by analyzing whether a layer contains pollen of tropical or arctic plants (paleo-climatology). Pollen analysis can also shed

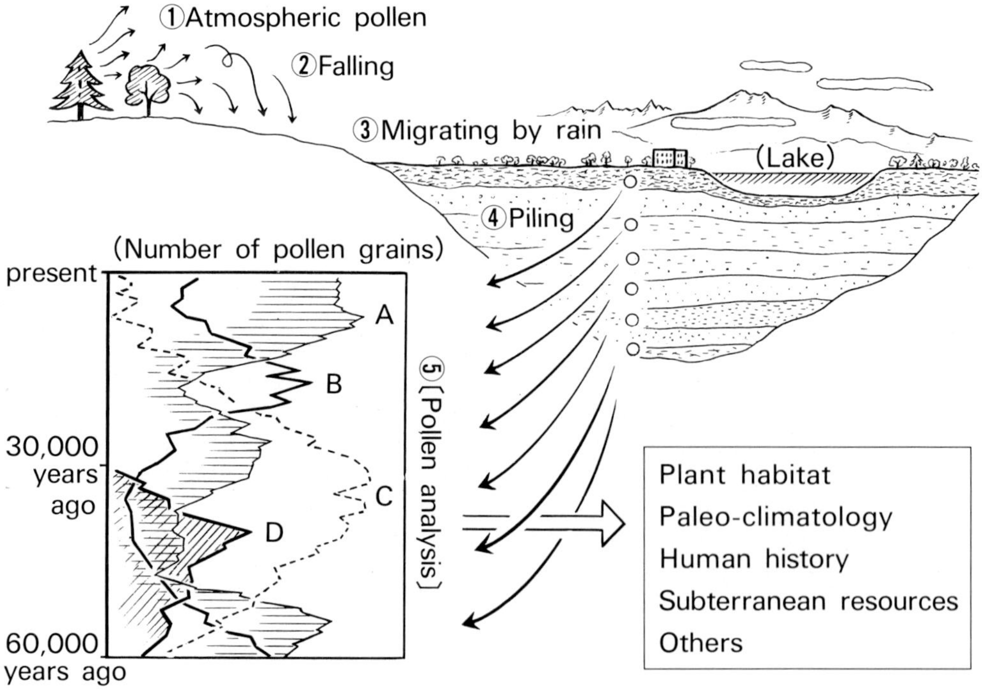

Fig. 7.1 Pollen analysis revealing various events in the past.

light on human history in prehistoric times, e.g. the migration of a human tribes and agricultural history. It is also useful for discovering and developing subterranean resources (Fig. 7.1).

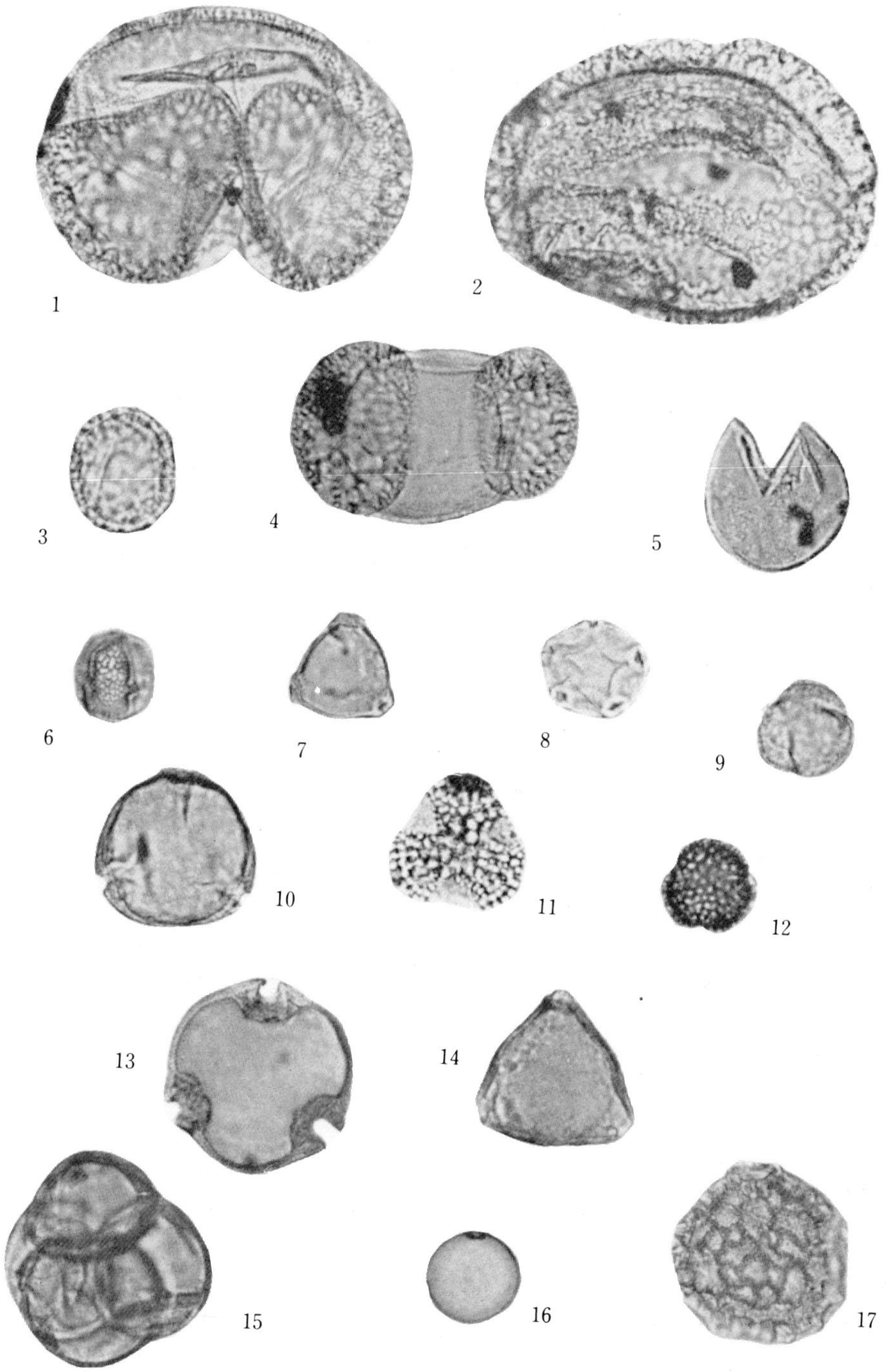

Fig. 7.2

1. *Picea;* 2. *Tsuga;* 3. *Sciadopitys;* 4. *Pinus;* 5. *Cryptomeria;* 6. *Salix;* 7. *Betula;* 8. *Alunus;* 9. *Quercus;* 10. *Fagus;* 11. *Ilex;* 12. *Ligustrum;* 13. *Tiha;* 14. *Elaeagunus;* 15. *Ericaceae;* 16. *Graimneae;* 17. *Persicaria.* (Yamanoi, 1970).

7.2 Methods of pollen analysis

Although it varies by type of sample (peat, coal or rock), a generalized method of pollen analysis is illustrated in Fig. 7.3. Carbonized samples must be prepared with both alkali and acid for pollen observation, but those from layers from relatively recent days such as peat after the Quaternary period can be prepared with only alkali for mounting on glass slide. Older rocks containing pollen grains or spores must be treated with 60% fluoric acid for about two hours.

Figure 7.4 shows pollen grains of *Aquilapollenites*, which does not exist at present, isolated from a soil layer of the Cretaceous period indicating that the species propagated the earth a hundred million years ago.

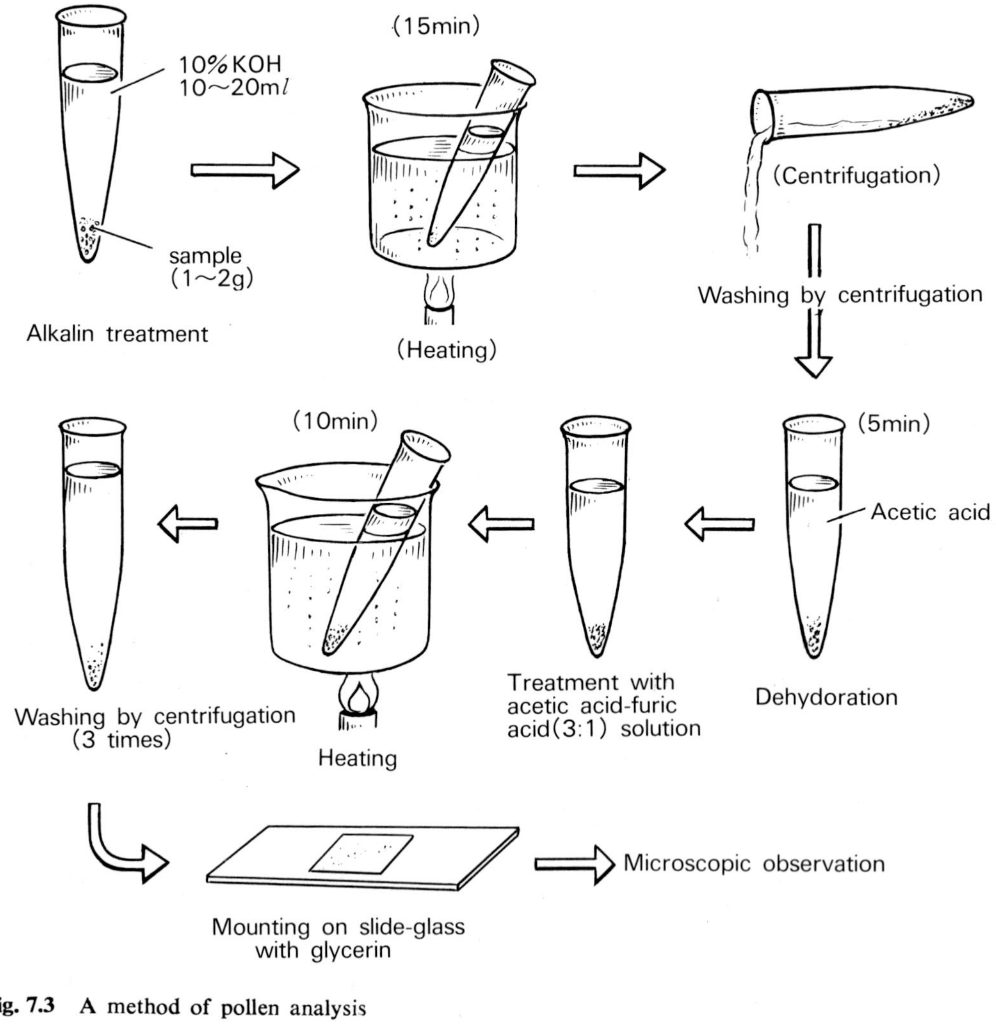

Fig. 7.3 A method of pollen analysis

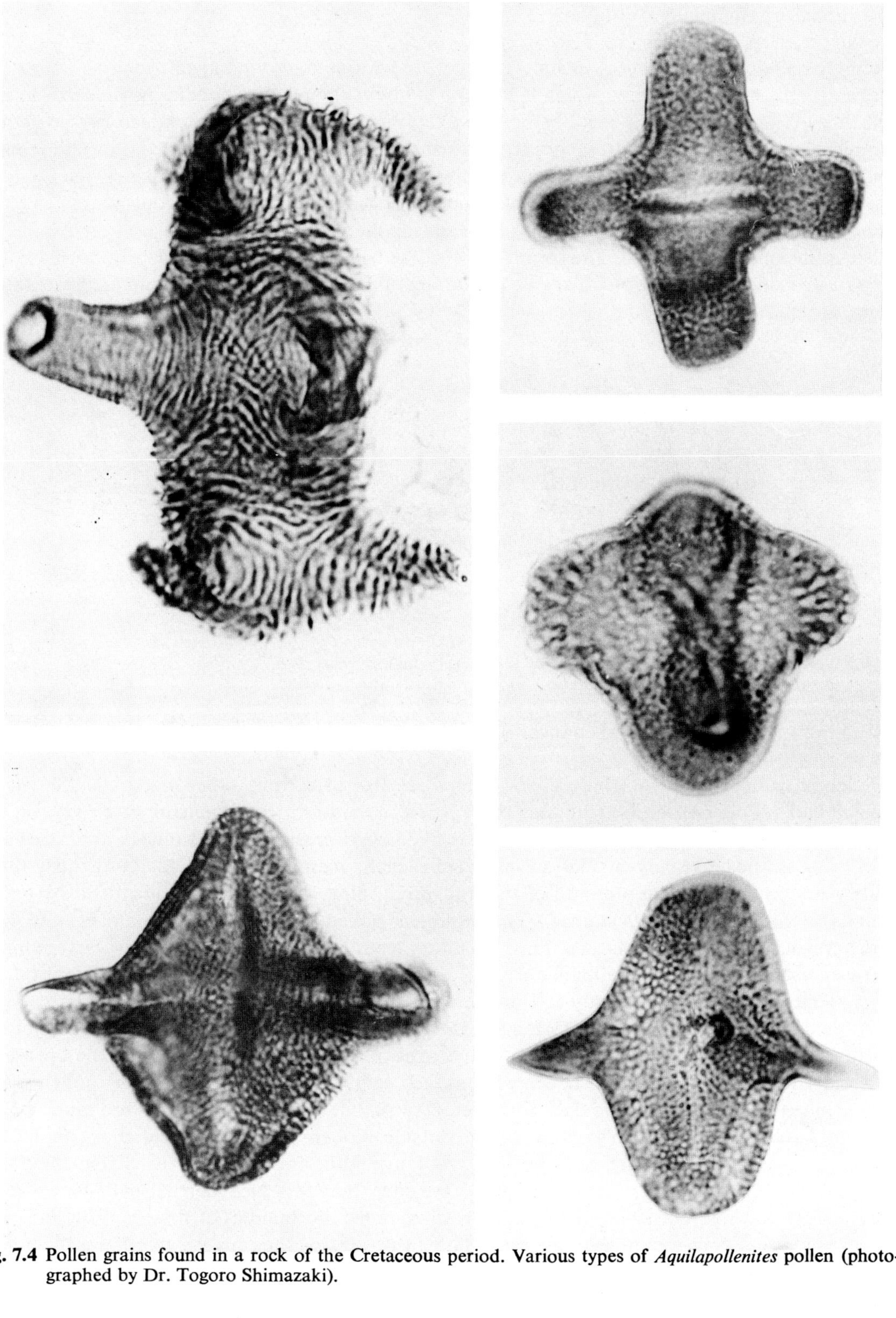

Fig. 7.4 Pollen grains found in a rock of the Cretaceous period. Various types of *Aquilapollenites* pollen (photographed by Dr. Togoro Shimazaki).

7.3 Microorganisms in soil layers

Besides pollen, the soil layers include residues of various microorganisms such as spores of moss and ferns or foraminiferas which tell us much about the organic conditions on earth in ancient times. The history of the earth before the Cretaceous period can be explained by the spores and microorganisms of spermatophytes (fruit baring plants), which have existed on earth from a hundred million years ago. Many reports are presented on these organisms in the pollen analysis session of the International Palynological Conference which is held every four years.

The spores, whose outer walls also consist of sporopollenin, of these species are very similar to pollen grains in size and shape. The spores are classified into two types, monolete and trilete; the former has a single germination furrow on the surface, and the latter has three. Some representative spores are shown in Fig. 7.5.

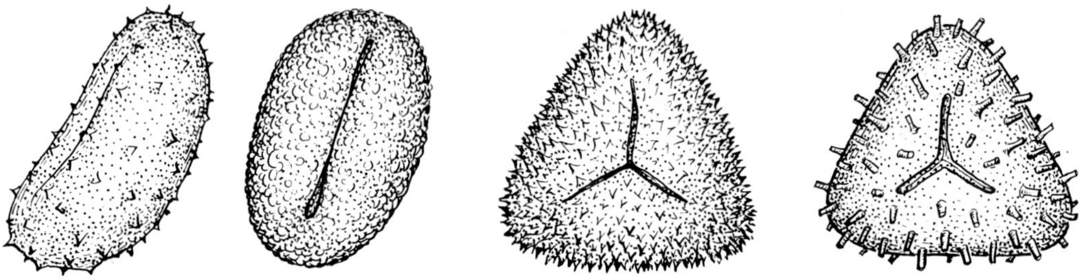

(A) (B)

Fig. 7.5 Various spores found from strata of a hundred million years ago. A: mono-furrow spores (monolete). B: tri-furrow spores (trilete).

7.4 Mammoth, Neanderthal man and pollen

Pollen grains are found not only in soil layers but also in various other materials. In 1948, a Russian North Pole Expedition team dug out a frozen mammoth from the bank of a rivers on the Taimyr Peninsula. The mammoth's mouth was filled with grass, and its stomach contained 200 kg of food. Radioactive carbon analysis indicated that the mammoth had died about thirty thousands years previously. The presence of timothy, birch, pine, and willow was identified by pollen analysis on the materials in its stomach. Among them, pollen of timothy was dominant, and 40% of the pollen grains were immature. The expedition team concluded as follows: In early summer thirty thousands years ago, the mammoth was eating timothy grass which was abundant on the Taimyr Peninsula, and it suddenly fell into an ice crack".

Dr. Solecki and his colleagues of Columbia University reported the results of the excavating the remains of Neanderthal man in the Shanydar Caves of Iraq in 1971. They found the bornes of a Neanderthal man, estimated to be sixty thousands years old, around which many pollen grains were identified. The cave was too dark for green plant habitation, and there was too much pollen for it to have been drifted by wind from the outside. Pollen analysis clarified more than eight species, including yarrow, bachelor's button, grape hyacinth, and Mexican tea. From the pollen around the bones, the investigators proposed the idea that this Neanderthal man had died in May or July and had been buried in the cave surrounded by bundles of flowers. This indicates that Neanderthal man was not an ape but a human possessing the capacity to admire the beauty of flowers.

7.5 Pollen a hundred thousand years in the future

Since pollen is useful for determining the existence and nature of subterranean resources, pollen analyses are conducted extensively in the USA and USSR on rocks in layers a hundred thousand years old. As mentioned above pollen analysis is based on the stability of pollen extine which consists of sporopollenin. The plants growing today also produce pollen coated with the same substance and will surely continue to exist in the world of a hundred thousand years from now.

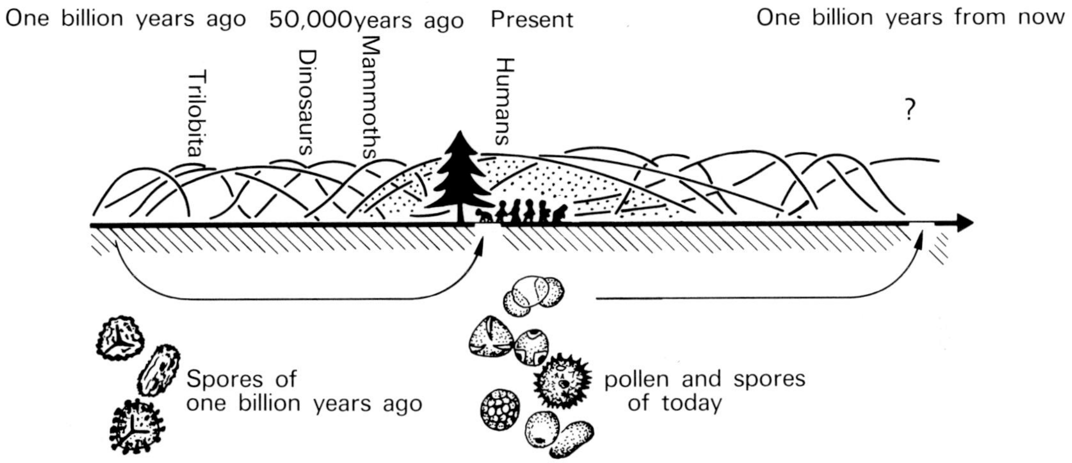

Fig. 7.6 Who will see the pollen grains of the plants of today a hundred million years from now?

Throughout the history of biological evolution, many species appear and disappear. Trilobita, dinosaurs and mammoths have come and gone. Who will analyze the pollen grains of the present species a hundred thousand years from now? *Homo sapiens* will probably not be in existence then. What kinds of creatures will these pollen grains encounter after we humans have disappeared off the face of the earth?

Bibliography

Brooks, J., Grant, P. R., Muir, M., van Gijzel, P. and Shaw, G., Eds. 1971. *Sporopollenin*. Academic Press: London.

Erdtman, G. 1954. *An Introduction to Pollen Analysis*. Chronica Botanica Co.: Mass.

Erdtman, G. 1966. *Pollen Morphology and Plant Taxonomy*. An Introduction to Palynology I. Angiosperms. 2nd ed. Hafner Publishing: New York.

Esau, K. 1977. *Anatomy of Seed Plants*. 2nd ed., Wiley: New York.

Faegri, K. and Iversen, J. Waterbolk, H.T., 1964. *Textbook of Pollen Analysis*. Hafner Publishing: New York.

Ferguson, I. K. and Müller, J., Eds. 1976. *The evolutionary significance of the exine*. Linn. Soc. Symp. Ser. No. 1, Academic Press: London.

Ikuse, M. 1956. *Pollen of Japansese Plants* (in Japanese). Hirokawa: Tokyo.

Iwanami, Y. 1980. *Pollen Biology* (in Japanese). Kodansha: Tokyo.

Knox, R. B. 1979. *Pollen and allergy*. Edward Anold: London.

Linskens, H. F., Ed. 1964. *Pollen physiology and fertilization*. North-Holland Publishing: Amsterdam.

Lyngheim, L. and Scagnetti, J. 1979. *Bee pollen: nature's miracle health food*. Wilshire Book: Calif.

Punt, W. 1976. *The Northwest European Pollen Flora I*. Elsevier: Amsterdam.

Punt, W. and Clarke, G.C.S., Eds. 1980. *The Northwest European Pollen Flora II*. Elsevier: Amsterdam.

Punt, W. and Clarke, G.C.S., Eds. 1981. *The Northwest European Pollen Flora III*. Elsevier: Amsterdam.

Roth, A. Ed. 1978. *Allergy in the world: a guide for physicians and travellers*. Univ. Press of Hawaii: Honolulu.

Stanley, R. G. and Linskens H. F. 1974. *Pollen: biology, biochemistry, management*. Springer-Verlag: Berlin, New York.

Ueno, J. 1978. *Study of Polynology* (in Japanese). Kasama: Tokyo.

Wodehouse, R. P. 1935. *Pollen Grains: their structure, identification and significance in science and medicine*. McGraw-Hill: New York.

Index